A VISION *of* HUMAN UNDERSTANDING

A VISION *of* HUMAN UNDERSTANDING

Emergent Philosophical-Scientific Paradigm

JAIME PRADILLA-SORZANO, Ph.D.
TRANSLATED BY RICARDO PRADILLA, M.SC., P.E.

CONTENTS

PART THREE

The Interior or Spiritual World

LIST OF TABLES

LIST OF FIGURES

PROLOGUE

Shortly before completing this book, which in itself can never be finished, I had the fortune of reading Karl Jaspers's philosophical essays called *Initiation to the Philosophical Method*, which made me understand once more that philosophical essence can be symbolized with the ellipsis, indicating that its dynamic character doesn't stop at any level of knowledge. Its truth is based on the sincerity in seeking it.

Jaspers, an existentialist, wished to bring to light, to the post-war German audience, the importance of philosophy, not only for its intrinsic value but for its method, which seeks knowledge without believing in certainty. The philosophical attitude gives man an approach that values his existence and makes him have his own opinions, away from the fashionable simplicities, tagged as rescue ideologies that pretend with their "novelty" to stifle liberty and thought. Similarly, it seems to me that the ideas expressed in this book, without pretending, of course, to have encyclopedic value, seek to provide a universal vision of knowledge in its multiple phases, in a dimension that reflects its historic evolution, favoring its comprehension.

In recent times, the evolution of knowledge has taken on unprecedented dynamic characteristics. Its frontiers are expanding rapidly, paradoxically giving new generations a sensation of frustration in their efforts for acquiring orientation that will facilitate their rise in the current complex world. On the other hand, the prevailing educational focus that has a marked pragmatic character, with emphasis on economical organization, leads to specializations in areas restricted to seeking physical sustenance.

Philosophical disciplines are considered "useless knowledge," for allegedly failing to provide the means to satisfy our "well being." Nonetheless, this observation is inconsistent, because its meaning depends on considerations that define the reach or context of the idea of "well being," and therefore implies making "value judgments," about which unanimous criteria would be hardly established, simply because they depend on our desires that are of a private character.

In the natural or physical sciences, the process for the acquisition of knowledge is conclusive. Its base is empirical; events are ostensive knowledge; methodology belongs to mathematical logic. In contrast, in "spiritual sciences," or, if it is preferred, "social sciences" such as history, politics, and economics, the relationship of subject-object turns on itself, becoming a self-analysis that necessarily introduces value judgments, which are of our own choice and freedom.

Following these ideas, we think they return to perennial philosophy, leaving us at its origins while simultaneously extending its current boundaries. We wish to give the reader of this book the opportunity to assess his or her own views regarding a wide range of perennial issues currently in dialogue with outstanding scientists, philosophers, mystics, writers, and historians who have been distinguished by their important contributions to human thought characterized by their coherence within their diversity. At the end of this book is a select bibliography representing the text, which is considered adequate to expand the reader's space of reflection. A good number of these books have been suggested and commented upon by my sons, Emilio and Ricardo, each one in their area of preference: philosophy and science.

The order in which these ideas are presented is not arbitrary; however, their continuity does not follow strictly a sequence in the way of a didactic text, allowing the reader to elect subjects of his or her immediate interest in an independent manner. On the other hand, it has been considered convenient, to facilitate reading of the main text, to point out some developments that by their mathematical nature, even though they contribute toward clarifying important scientific aspects, are not considered essential for general comprehension.

It is thought convenient to approach the significance of knowledge, reflecting on its historical origin, as much from its individual development as from its social evolution. Only by going back imaginatively to the early stages of our existence, particularly observing the marvelous dynamic that allows a young child to advance in a totally unknown world, can we begin to discern the meaning of our existence. The childhood educator, the mother who is so close to the beginning of life, has the occasion to contemplate this marvel of nature. My autumnal age has allowed me to return, along with my young son Juan Diego, to that transcendental era of our existence, to capture some of those impressions in this book.

Following the intellectual ascent of man, we value the importance of social interaction that contributes in a decisive way, through the symbolism of language, to cross the threshold of private knowledge toward public knowledge, which differentiates man from his congenerous fellows, all other animals. The recognition of this evolutionary trail leads to the analysis of language, where there is printed rational development that is made explicit by logic and mathematics. Since Socrates, it has been considered that philosophy originates in the analysis of language, which deepens into the meaning of words and their interactions. For the Greeks, philosophy encompassed all of knowledge; the widening of its boundaries during the Renaissance, primarily with the development of the physical or natural sciences, transformed Western thought with a new methodology that conjugates Greek rationalism and observation.

A broad view of the significance of our existence can only be reached by opening the windows of time, enabling us to observe a perspective of the path covered, valuing modern limitations. These ideas are framed in the philosophy of Bertrand Russell, who graphically describes them with the term "provincialism," which is generally understood as the product of a geographical limitation of our observation, forgetting frequently that we are "provincial in time," remaining confined in the instant of our era and our self. I consider that Russell's rationalism and my enthusiasm for scientific knowledge shaped my youth and my intellectual evolution, moving further away from the context of Colombian culture at the beginnings of the twentieth century, which oscillated between scholastic philosophy and the French encyclopedists. With maturity, my admiration for the physical sciences was tempered as I recognized their limitations, moving toward an appreciation of "perennial philosophy," as described by the English novelist Aldous Huxley, who recognized the spiritual experiences that give a sense to our existence.

These phases of my intellectual development are reflected in the way of a holographic presentation of this "vision of understanding," which aims to balance the rationalistic tendencies of the scientific method, based on the so-called "objectivity principle," with the observation of our inner world that values our feelings and desires.

It is my hope that the reader's desire for knowledge will awaken his or her intellectual curiosity, signaling simultaneously immediate and global aspects of the most relevant ideas expressed in this synthesis of the intellect. In conventional wisdom, it is often said that we can't see the forest for the trees. Well then, to see the forest it is necessary to dive into its foliage and observe its beauty, its dangers, and its diversity. However, we must not stay in its environment; we must scale the mountain and contemplate its expanse, its magnificence, and its relation with immensity and our own selves.

One of the marvelous characteristics of the mind is its capacity for global comprehension, preserving its sharpness and ingenuity. In the ideas expressed in this book, we emphasize a holographic vision of knowledge, which necessarily requires us to scrutinize and analyze specific aspects of our observations, so desirable for evaluating the whole.

In the first chapters, as we analyze the origin of thought, noticing the diversity of our perception, we try to explain it with symbolism, inquiring in the fountain of rationality, the language. If we analyze carefully the structure of that symbolism and we try to train ourselves in its algebra, we must understand as well, that its rigor is not essential for ascending to higher studies. In a similar way, to comprehend the development of physical science, it is necessary to assess the importance of mathematics in the logical development that relates the hypothesis or postulates with the observations of events. Moreover, our understanding should progress regardless of detail, at a personal rate and rhythm, to return if we wish extending our vision.

Following this direction as we study the mechanics of Isaac Newton, we can comprehend its ingenuity without sharing the mathematical terms of infinitesimal calculus, understanding the significance of its postulates and its importance in the development of science in general, and in particular for the establishment of classical mechanics and further improvement through the ideas of relativity.

Comprehension of the scientific method, via its historical description, points out its evolutionary features characteristic of the natural sciences, which are manifested according to the complexity of the system to which they refer. The principles of classical mechanics have the spirit of Aristotelian logic and the methodology of Euclidean geometry. Relativistic mechanics expands its postulates, refining its logical rigor as it advances toward the principle of objectivity.

In a more complex realm, such as chemistry, its description and evolution is initially empiric and its concepts are of greater mathematical simplicity. Newton's mechanics didn't have the necessary elements to formulate a theory for material diversity. The ingenuity and experimentation of the new alchemists made possible the establishment of basic atomic theory, which in turn allowed Newton's mechanics to be moved to the micro cosmos, facilitating mathematical formulation of the new physical-chemical science. At this stage, some phenomena were discovered that could not be explained by classical mechanics, giving place to a mathematical reformulation called quantum mechanics.

In that section of the book, a detailed description will be found of the main experiences that defined the atomic model. Because the theory of quantum mechanics is of a mathematical nature, its concepts require the use of certain concepts from wave theory and electromagnetism, which are presented in a didactic manner, separating the principal text from some mathematical developments that, even though they provide greater clarity, are not indispensable for comprehending the general thematic.

Biology is the new frontier of science; its presentation in this book has a general character that seeks its theoretical foundations. An attempt is made to analyze philosophical and scientific aspects that have great impact in the study of human sciences, such as Darwin's evolution, within an actual context in the scientific frontiers that moves away from determinism of traditional science by using novel mathematical "nonlinear" procedures applicable to complex systems.

In the last three chapters, we reflect on philosophical aspects of the significance of words such as *spirit* and *matter*, by the light of metaphysics and science, as well as on psychological concepts on the development of the "ego," compared with the spiritual transcendence of Eastern philosophical mysticism.

Ethics and politics contain philosophical and scientific aspects of the utmost importance to human welfare; the recognition that, by nature, they are so close to our individual and social interest takes them far from the realm of logic and science. The definition of values that give meaning to our existence includes an extreme range of concepts that fall within altruism and selfishness, idealism and empiricism, materialism and spiritualism.

The chapter on ethics and politics provides an orientation to this controversial analysis, which is consistent with the general comprehension of scientific knowledge and its methodology, as well as with abstract ideas, leading to a reevaluation of spiritual concepts.

Jaime Pradilla-Sorzano
February 2015

PART *One*

___

THE RATIONAL OR IDEAL WORLD

1

INTRODUCTION

1-1 SCIENCE AND DOGMATISM

When we reflect on ourselves, many more questions than answers arise. From an early age we question our own identity. Why do we have our own personality and not a different one? We always remember being ourselves and not somebody else. What does our individuality consist of? Is it real or is it fiction? Many of these questions of our youth become dormant in the vast majority of people, to give way to an empiricism imposed by culture as a pragmatic attitude, which should replace all those childhood dreams by clear and defined concepts that give us a strong foundation capable of facing a difficult world in which personal success is what is important.

This is how we go on shaping our beliefs, which are in their majority a product of the culture of the society in which we are developing. However, questioning is in our nature and, if stimulated, it will lead us to philosophy, a word which etymologically means love for knowledge (not for "sophy"). It is the discipline of the wisdom that supports the analysis of our ideas and our knowledge; its principal characteristic is that its answers are never definitive, always producing new concepts and questions that may go on indefinitely and in fact are a part of our own existence.

In opposition to this philosophical approach is the dogmatic approach, which takes for granted as irrefutable that some concepts we must accept as fully established by some magical mechanism of our traditional culture.

Unfortunately, religion has been established historically through a mythical mechanism, which implies in a greater or smaller degree a dogmatic position. The confrontation between this religious position and that of the natural sciences, which has a critical and philosophical position, has brought many to relate religion with the Dark Ages and its prevailing widespread ignorance.

Modern science has given mankind an extraordinary power never before imagined. As such it can make us self-destructive or lead to a stable well-being expandable to the great majority of our population. As a consequence, in our current society, science's following has its origin almost exclusively in a pragmatic understanding based on its extraordinary results of dominion and control over nature rather than on its intellectual values, which are largely unknown to the great masses—including some of science's most avid admirers. This fact has propelled new beliefs that, dismissing the origin of natural sciences, generalize its scope into realms where it does not apply. On the other hand, it is philosophy's task, with its more general and fundamental approach, to dispel these neo-dogmatisms that threaten once again with their academic obscurantism. Mankind, during its cultural evolution, has narrowly escaped from its own egocentrism, giving way to great ideological transformation, such as modern science, which has been one of its principal achievements. Nevertheless, science's extraordinary results of dominion and control over nature have generated new dogmatic beliefs that contradict the very origin of natural sciences, which has based its critical position on new and open concepts.

As parochial as a local villager may be when he reduces his world to his village, is how a "wise man" may be within his pompous science and arrogance, thinking that his small world is the truth unveiled. In addition, we simple mortals, who are not part of the "scientific nobility" and are not registered in its "jet set," may easily fall victim to this narrow, intellectual parochial view of our time, directed by our media, educational systems currently in vogue, and the technological civilization that worships a new god—the pseudo-science.

Scientific comprehension is based on the confrontation of experimental observations of an outside world, which we assume can be organized in a similar fashion to our mental events previously and intuitively conceptualized. Frequently our desire to extend these processes of physical and mathematical sciences outside of their real dimensions leads us to extrapolate their validity to realms that are not appropriate.

A new synthesis is pointing our way today, which appears to be a time of crisis. How can we contribute to the search that will consolidate this new cultural direction? This project will define the quality of life for future generations. Its message is a vision of our knowledge within a historical perspective, using this word *knowledge* in its broadest form, analyzing human thought in a multidimensional space, and liberating us from the contemporary attitude that is polarized by a pragmatic and excessively individualistic view that only values economic acquisitions and power.

1-2 OUTLOOK AND APPROACHES OF KNOWLEDGE

It would seem redundant to refer to knowledge as only human; apparently we don't know any knowledge that is not. In reality, this attitude only shows how pride can lead us to the belief that our understanding is isolated and unique with respect to our planet's life.

We may analyze knowledge from different points of view, according to its reliability or intrinsic value, its usefulness for an individual or the species, its historical development, or its cosmic transcendence.

If we try to evaluate our concepts or basic beliefs that sustain human knowledge, we can opt for different analytical views, which assign them labels such as philosophical, psychological, or physical-mathematical. It is not an easy task to define these fields of thought, and if we attempt it, we will realize that their environments overlap and interact. In turn, each discipline has subdivisions—for example, philosophy may include logic and metaphysics, and psychology may be behavioral, experimental, or refer to a branch based on introspection. In physics, which is more compact, the divisions are subtler and have an origin mainly based on their historical development.

How can we group these multiple disciplines of knowledge in a way that allows us to perceive basic, interrelated entities, while preserving their identities? Are there any truths independent of our existence that have a validity outside of time, our mind, and the physical world? Are the ideas of logic and mathematics, the eternal truths of the world of Plato, preexistent to the physical world? Or are they simply a creation of our mind, which in turn is an indissoluble part of nature?

Let us consider that these entities, designated by Roger Penrose[1] as the Mathematical, Physical, and Mental Worlds, properly described in *The Road of Reality*, which for our purposes, facilitates the division of the content of this book into three parts:

First	Rational or Ideal World
Second	Physical or External World
Third	Spiritual or Interior World

[1] Roger Penrose, *The Road to Reality* (New York: Random House, 2004).

Emphasis on the analysis of acquisition of knowledge, which we may call its "history," is one of the ways that may contribute most to its valuation. We may observe this "history" from an individual point of view, from a species view, or in its cosmic origin. The natural sciences and psychology implicitly include the history of human knowledge; we only need to elucidate the subtle outline of relationships and patterns from the list of events and facts that constitute the empirical data that form the scientific basis. This reflexive or philosophical labor allows us to assess the reliability of our knowledge, and gives us individually or collectively an ideology or a paradigm that replaces the traditional myths that we have inherited.

It is precisely this historical point of view that will guide us with higher certainty to an evaluation of our beliefs. History is the key; everything is understood when we know how it was formed, what is its origin, and how it developed. Frequently, an elusive abstract theory stops being difficult when we reduce it to its basic concepts, which are printed in its historical development.

Our analysis of thought should be reoriented toward the origins of life or the cosmos; this view is undoubtedly fruitful to situate us in an environment that will break through our temporary limitations, reflecting cosmic history on the evolution of every being from a rock to a human. Considerations on the development of a human being, from the fertilized egg to the mature adult, reflect in its different stages the history of the human knowledge. The development of a child's body in the uterus recalls its biological history, from the origins of a single and primitive living cell to the appearance of the *Homo sapiens* on the face of the earth.

Considerations about life—its biological and evolutionary history in general and its cultural development through different species including humans—give us a particular perspective of our own value as a species and as individuals. The universe and its development lead us to the confines of space-time and give us an almost infinite perspective, integrating us within an order of all the events at all the times and places. Perennial philosophy, which emerges from its own history, shows us that ideological plurality is the source of knowledge and its evolution, in opposition to the paralyzing dogma.

A child's vision and development, mainly in the early years, show the basic meaning of our thought. We will understand the adult mind better if we analyze its evolution. How we have arrived at the intuitive concepts of time and space will be a firm base to understand their validity and their limitations. Likewise, the origin of human language will give us very valuable information on philosophical concepts of mysterious appearance such as matter and spirit, which have a great respect attached to them, so they can be reduced to their original simplicity, stripping them of such pompous significance as our cultural evolution has assigned to them.

The events and their order, which have shaped our existence, are rooted in our minds. This register is a basic part of the ego, to the extent that it is confused with our own identity. The content of this memory is not simply a list of data; its order defines to us time and space, and its relationships take us to the logical concepts of cause and effect.

Physical-mathematical concepts, as abstract as they may seem to our naïve appreciation of the world, can be traced to their true universal dimension, if we look for their origin and historical evolution, from an intuitive point of view, trying to glimpse the original mental processes that conferred them their validity.

2

FIRST VISION OF THE CHILD

We appear unexpectedly in an unknown world, without reference, because we don't know any other; we also arrive at the last moment. We cannot even ask any questions, because we don't know how. Nothing is new or old, time has not even begun or it has always existed. We come from the empty space to the material space and this is our legacy. What to do with it? Herewith is our life; we have to develop it. It is a titanic challenge.

Is everything solitude? We are already attending the apostolic mass, and we don't know who Jesus is, why he says good-bye. Where is he going? What language does he speak? What does this all mean? Nobody has invited us, not to any given place or time, not for any reason. We have arrived late and it will always be this way. Nevertheless, we dispose of the book of the history, our mind, which is the order of the universe. Within the contrast of that external world and our own lays the meaning of the life.

Our mind is a blank book that has subtle patterns that guide us in the right direction. Those invisible guides are our history, which reflects the order of the universe as a whole in our mind. It is our heritage, without which our labor would be impossible. This is the history of humanity, life, planet Earth, and the universe. To perceive the contrast—between light and darkness, heat and cold, silence and noise, stillness and movement—is the only key to begin our search.

Some simultaneous sensations give us clear impressions, like black and white, and they may suggest for the first time the position of "something," and from there on we pass to the multiplicity, several "somethings," and we then begin thinking about space. Other sensations have to be repetitive to form a habit; memory allows us to compare or contrast sensations and initiates us into the time dimension.

The concepts of space and time, as the feeling of consciousness or individuality, begin their process of mental development from early infantile sensations. The distinction of different blotches of colors in the visual field gives a first idea of position, which is complemented by sensations of displacement of our body or a part of our body, like a hand or a foot that touches something. Repetition of these sensations makes us form little by little our intuitive or psychological concept of space.

Simultaneously we begin our physical individualization, when we distinguish between our body and other external sensations. This single physical feeling is the first step in the formation of our ego. Animals also reach this stage of individualization, but at a much earlier stage in life than we do. We believe that their concept of ego doesn't go further than this level because its development in man is shaped in good part by means of communication through language. However, from the evolutionary point of view of life in general, we can believe without a doubt that the physical ego is present in the phenomenon of life starting from one-cell organisms that through evolution have established continuity up to humankind.

To analyze the first sensations of a child, we cannot appeal to our own memory; only by extrapolating our present and placing ourselves in the situation of a child's mental emptiness, will we have a basis to evaluate the initial development of the infantile mind. Our ingenuity will play an important role in connection with the observations of his or her behavior.

Therefore, we may conclude that a good part of the psychological theories based on introspection are imaginative, and by their nature cannot be expected to be confirmed objectively as happens in the natural sciences, especially in physics, which uses mathematical symbolism.

The basic information that we receive through our senses is a contrast or comparison of signals. A child begins his or her mental development distinguishing between light and darkness, silence and sound, heat and cold, touch and vacuum. This essentially binary communication (yes, no) is the fundamental language of electronic computers, and is inappropriately named as "digital." In machines this information is registered as "bits of memory" in the crystallographic grid of "chips," which acquires an order that can be interpreted as a number in binary code, employing the signs 0 and 1.

The complex processes related with our biological memory are not known sufficiently at a physical-chemical level, but without a doubt significant progress in this field is expected, as well as in the field of identification of cerebral areas and their interrelationships with memory. Nevertheless, as important as the theoretical and practical consequences derived of these correlations among physical and mental data may be, these methods will not substitute for the introspective analysis of thought, but rather will complement it.

On the other hand, the psychological mechanisms that characterize the memory of human beings tell us that they are of a type very different than the memory of computers. It is a very well-known fact of our daily experiences that our punctual memory, that is to say the one that refers to something specific, such as the memory of a distant person from our past, may appear unexpectedly triggered by association with a stimulus like a musical tune. Also suggesting a dynamic system of our memory, are the experiences of psychologists and psychiatrists such as Daniel L. Schacter,[1] who tells us:

> The complete rehabilitation of amnesiac patients, who have remained during long periods amnesiac by reason of traumatism, recover their personality in surprising form, and their memories come back like a video stimulated by the repetition of a past occurrence that when being revived, releases a torrent of images and feelings that make them to feel once again as the beings they once were.

Human memory is simultaneously global and punctual, in contrast with the "memory" of computers, which sequentially passes from one piece of data to the following one. This suggests that neural networks operate through communication mechanisms very different than electronic wiring. It is known that the plasticity of neural communications confers them mobility through space and time; however, this wonderful property doesn't seem enough to explain our consciousness as a global phenomenon of our memory and understanding.

The search for a physical explanation of these global phenomena of communication within our minds suggests a quantum delocalization of these systems. This mechanism would be able to establish a resonance, that is to say a synchronous collective neural phenomenon involving a psychic field associated with our consciousness, which in itself is considered a global state of understanding, stimulated by a punctual event. This is the physical reality that suggests a psychic event, as the complete rehabilitation of an amnesiac patient in such unexpected fashion.

---

[1] Daniel L. Schacter, *Searching for Memory: The Brain, the Mind, and the Past* (New York: Basic Books, 1996).

That sense of being, known as consciousness, has many facets, starting from physics as the most elementary, to intellectual understanding, which allows us to find abstract knowledge, a characteristic of philosophy, mathematics, science, artistic expression, and transcendent mysticism.

The concept of time, which we may call psychological, is complementary to the sensation of space. Its gestation is more elaborate and is implicated with the sense of ego, with our consciousness that is manifested when our memory is global, with a simultaneous integration of our experiences. Psychological time only occurs with a psychic state that we call conscious. On the contrary, in physical science, time is an "external" concept we relate with events that happen independently of our consciousness. Partly, the main difficulty in understanding scientific theories such as relativity is derived from trying to reduce these two concepts into one. Psychological time is private; physical time is public. We shall have occasion to delve into these important differences when we look at large private and public events. In an electronic computer, linear sequences of events that may be related to each other can be registered, allowing physical time to be established; this is not the case with psychological time, which refers to oneself only.

A child develops initially the concept of psychological time simultaneously with his or her individual consciousness. When the child is hungry, separated from the mother, or has a painful sensation of hot or cold, these feelings individualize his or her consciousness of the passing of time, which happens quickly or not at all when the child is distracted, and is not aware of the self. This primitive concept that we have called psychological time presumably in the early stages of development is also shared by superior animals, which undoubtedly manifest a primitive individual consciousness.

Knowledge has two basic elements in its development: the simple recording of sensations or data, and their ordering, which seems to take place simultaneously in a nonlinear or nonsequential fashion, forming multiple relationships that we can visualize in a multidimensional network that is characterized by various stages of communication, which follow different levels of order.

When a series of sensations is repeated, the sensations are recorded in different sequential states that form an order in the registered events within our memory. Undoubtedly the slow acquisition of the concept of time is a more complex process than that of the space concept. The simple memory of data divides the field of consciousness in two: our current sensations and those already registered. We begin to differentiate present and past. To go further it is necessary to sort the acquired data sequentially. How do we do this? It seems that the association with other sequences is the only way to initiate this process. We are faced with the quintessential question that makes children smile: Which came first, the chicken or the egg? How do we get out of this dilemma? The solution leads to the theory of the evolution of life, where the sequence egg-then-chicken or vice versa is part of another wider mechanism, in the evolution of one-cell organisms to multicellular ones.

Are the elementary concepts of time, space, and individuality already printed in our brains? The evolutionary process of life has left subtle patterns that reflect the order of the natural environment in which it has developed. We are not so alone; the life of each one is a continuum. A newborn has the experience acquired by life in its millions of years of interaction with the environment.

Returning to our question of how to organize memory, we can imagine that a basic sequence of order already exists in our mind, without resorting to external data. It seems reasonable to assume that those ancestral patterns give us the ability to manage early visual and auditory scales of feelings or of another order, which may be related to each other, establishing a base for memory of more complex life experiences.

Undoubtedly, in animals, we don't know up to what level this capacity is printed already from birth, which allows them in a surprisingly quick way to develop the appropriate abilities to interact with their environment in a short period of learning. Contrary to popular belief, other living beings besides humans use elementary concepts of time, space, and physical awareness of their individuality, indispensable for their existence. To what extent does mankind differ in the establishment of these fundamental concepts? It is believed that the invention of symbolism that we call language made possible such a discontinuity in human evolution.

3

SYMBOLS, ORDER, AND LANGUAGE

Order is a very particular concept; all people have their own, within their mind and their immediate physical domain. If somebody comes to change the way we collect our documents, pictures, or goods, we are immediately startled because the "disorder" imposed on us is not our own.

Humanity began its cultural ascent with the invention of symbols that are the basis for internal and external communication. The organization of data that we receive from the outside world through our senses is carried out by means of symbols that may have an internal validity, which is private, or by means of external communication through language, which, in addition to including words or sounds, is also modulated by body language, inflections, and tones of the voice.

Human evolution in its most recent stage, as it definitively moved away from the other species, was carried out by means of symbolism that enables communication, which is the essence of the thought process required for individual, as well as social or cultural, development. In abstract terms of communication that are used in electronic information technology, everyday language can be considered "digital" lingo when it is expressed by means of letters or numbers, or "analog" when drawings or corporal symbols are used. Primitive languages are closer to the actions they describe with their onomatopoeic sounds or analog, and their written symbols are pictorial or digital, as in Chinese writings.

An esoteric aspect of a language's symbolism is its character of internal communication in our minds. Its evidence is proved in our experience that communication, especially spoken, stimulates thought. Frequently, as we speak our understanding is illuminated and new ideas appear. A professor knows very well that to teach is the best way of learning. It is important to understand the significance of language or symbolism in our mind's formation, not only from the cultural point of view, but also individually—and hence in the evolution of the human species.

The concept of order is built by a correspondence between our mental system and a group of events or things that we designate as external. That relationship, if it is conscious, becomes expressible by means of language; however, this has subconscious and intuitive elements, as those found in mathematical logic. For this reason, analysis of language has been, since Socrates at least, one of the means to discover its latent philosophical fundamentals. Logical symbolism extracts the implicit relationships found in common language, expressing them in a general form, in a similar way as does algebra by replacing numbers with letters to express their relationships by means of equations or algorithms.

3-1 MATHEMATICAL LOGIC

Historically, logic in the West has its origins in Greek philosophy and mathematics, known in medieval Europe through Aristotle and Euclid. Algebra, as indicated in the word's etymology, has its origin in the Arab culture. Currently, algebra is more widely diffused than geometry and logic, following the pragmatic tendency of modern education, which virtually suppressed the fundamentals of arithmetic and geometry and relegated philosophy and grammar.

The invention of Descartes's coordinates made possible the integration of geometry and arithmetic, and, at the end of the nineteenth century and the beginning of the twentieth century, symbolic logic emerged as the foundation of mathematics, identifying both disciplines as mathematical logic. The utilitarian approach of current culture has caused mathematics in general to be spread exclusively from the point of view of its application value, reducing its logical fundamentals to Greek syllogism, which Thomas of Aquino's scholastic wisdom undusted from Aristotle's teachings.

Our age is characterized by ignoring basic knowledge that is considered useless. This is an accelerated trend imposed by modern technology and the eager, indefinite growth of the economy, which aims at developing its own vicious circle as an end in itself and not as a means of well-being. This "fad" has even permeated into our intellectual class, the one that should be the elite that could maintain the study of the basic principles of knowledge. Political pressures have influenced scientific publications to place most of their emphasis on results, as hopefully or apparently utilitarian as they may seem, forgetting that development from the mental point of view has a far greater value than the transformation of the physical world.

The vast majority of people, even in the most advanced countries, ignore science as an intellectual value. Their admiration and respect is only reserved for its utilitarian aspect and therefore it is confused with technology. This attitude, which is also rooted in the so-called intellectual elite, unavoidably forces modern society toward a new "scientific" dogmatism that denies the conception that originated it.

Along these lines, and as a contribution to the exaltation of the intellect, let us briefly recreate our look into logic.[1] That requires looking into language for its deductive modality, which is called syntax in grammar, reducing it to compact symbols that facilitate operation in deductive reasoning.

Let us analyze some phrases or sayings that we can also call sentences or propositions, such as:

All - men - are mortal.   If - John eats - then - he gains weight.

All - flowers - are beautiful.   If - Peter laughs - then - he is having
                                           a good time.

In these phrases we can distinguish "factual elements" that refer to facts—such as *men, flowers, are mortal, flowers are beautiful, John eats, Peter laughs, gains weight, having a good time*—that can be substituted for others without changing the structure of the proposition. On the other hand, we find other phrases or words, such as *all, if,* and *then,* that cannot be substituted with others without modifying the structure of the sentence; these are called logical.

The previous propositions are grammatically correct and they can be true or false, whether or not they correspond to facts. The facts themselves are neither true nor false; they are occurrences: True fact is a redundancy and a false fact is a contradiction. A fact corresponds to two propositions, a true one and another false one.

Propositions can be simple (atomic) or compound (molecular). The atomic ones cannot be broken down and compound ones are constructed from more than one statement. In the given examples, "All flowers are beautiful" is atomic, as is "All men are mortal." The other two examples are a molecular proposition.

---

[1] J. Ferrater and H. Leblanc, *Lógica matemática* (Mexico: Fondo de Cultura Economica, 1994).

Two modalities are considered in logical analysis: propositional or sentential logic, which relates propositions to one another, and quantificational logic, to which we will refer later, which uses quantifiers like *all* or *some* in those relationships. In the first one, we relate the propositions by means of logical particles, such as:

or,   and,   if

which are symbolized as   V   •   >   respectively

and serve to build new propositions. This way, in our example, we can designate for the letters p,q,r the propositions in general: "Peter laughs" = p, "having a good time" = q, and the resulting proposition, "if Peter laughs then he is having a good time" = r. We express the relationship as:

$$(p > q) = r$$

In words this says: if p then q, we call it r. The truth or falsehood of the new proposition will really depend on the condition of truth of the propositions p, q that are related by means of the logical particle that it is used in this relationship. The result of some of these relationships are represented in the truth tables (Tables 1 and 2), where the columns present the conditions for truth (T) or falsehood (F) of the molecular propositions that result from relating the propositions (p, q) by means of logical symbols (V, ., >) or combining these in a logical formula, for all the possibilities of p and q.

p	q	p.q	pVq	p>q	p>(pVq)	((p>q.q>r))>(p>r)
T	T	T	T	T	T	T
F	T	F	T	T	T	T
T	F	F	T	F	T	T
F	F	F	F	T	T	T

Table 1: TRUTH TABLE OF PROPOSITIONS P, Q

We will find three cases in the truth tables; for each symbol (column):

1. Columns that only have T
2. Columns that only have F
3. Columns that have F and T

In each case we say that the corresponding relationship or formula is:

1. Tautological or necessary
2. Contradictory or impossible
3. Uncertain or possible

It is worth noting that the tables of the elementary logical signs (those that cannot be broken down into combinations of others) are not deducible by means of a computer because they are definitions; their interpretation is in the metalogic. However, the truth tables for compound symbols, as is shown in the last two columns, can be deduced by means of combinations of the values given in the tables for the elementary symbols. It will be enough to write a program replacing T=1 and F=0 and to make the appropriate substitutions successively to obtain the new column. This computability of the compound logical symbols shows us its deductive power, also a characteristic of mathematics. Besides, let's observe that the elementary or primitive symbols that are used in a given calculation may appear as defined in another schema and vice versa. For example, when introducing the negative symbol " – " we will be able to define (def.):

$$(p. q) = \text{def.} - (- p \lor -q) , \quad (p > q) = \text{def.} (- p \lor q)$$

leaving as the only primitive symbols "–" and "V".

In particular, tautologies will be of great utility, although their name indicates that they don't say anything new, but state what is obvious. We will see with some examples that the so-called principles or laws of classical logic are tautologies. It is worth referencing Ludwig Wittgenstein, who said: "Propositions show what they say; tautologies and contradictions show that they say nothing," and "Tautology and contradiction are not figures of reality, but are not absurd." We will return to these ideas in the next chapter, where we will analyze the knowledge of natural science and the mathematical-logical procedures that it uses in its inferences.

The following paragraphs referring to the foundations of logic and its origin in connection with language are not essential for understanding the continuity of ideas presented in this book, and the reader may omit them, if not interested in such technical matters. In a similar fashion we will indicate in other sections of this book some other writings that may not be needed to grasp the overall context of this book.

SOME LAWS OF TRADITIONAL LOGIC

Let us review some laws of traditional logic. We will use Table 2, which refers to a single proposition (p). The symbol "-" is denial.

p	p.p	pVp	p>p	-p	pV-p	-(p.-p)
v	v	v	v	f	v	v
f	f	f	v	v	v	v

TABLE 2: TRUTH TABLE OF A PROPOSITION, P

There are three tautologies in the table:
1. Identity Principle (p > p)
2. Contradiction Principle (p V -p)
3. Exclusion Principle - (p • -p)

Using language we would say:
1. If a thing is, it is.
2. A thing is or it is not.
3. A thing cannot be and not be.

Obviously, if we reduce these principles to everyday, infantile sentences, we see that they do deserve to be called tautologies, since they are unconditionally valid. But let us see if they can be of any use.

One of the objectives of logic is the construction of tests for a proposition or theorem. Proof in logic consists of obtaining from the premises the conclusion by means of inference rules. These rules are:

1. SEPARATION RULE

If a conditional proposition and its antecedent are taken as premises, the consequent result may be inferred as conclusion. Thus, schematically:

p > q} true

        (or premises)   → q (true consequent)

p } propositions

This should not be mistaken with the tautology ((p > q) . p) > q that is always a true proposition as a whole. The separation rule refers to q only. Keep in mind that the premises p > q, and q, do not deduce p. This rule seems to be obvious, even though obvious or intuitive must be established as such.

2. RULE OF UNION

If two propositions are taken as premises, then their conjunction can be taken as conclusion.

p

} premises → p • q conclusion

q

This rule is implicit in Table 1 in the first row of T with the respective columns.

3. INSERTION RULE

Any example of tautology can be taken as premise in any proof.

Proof example:

We want to prove that the following deduction is correct:

p > q

} premise → p > r as consequence.

q > r

a. We used rule 2 and united the premises, obtaining (p > q) . (q > r)

b. With rule 3 we used tautology (Table 1, last column) as a premise:

((p > q) . (q > r)) > (p > r)

c. With rule 1, taking the consequents of a and b like premises, we deduce p > r according to the scheme we wanted to prove.

Quantificational logic, also called functional, is characterized by such words as *all*, *none*, and *some*. Propositions are analyzed by distinguishing in their structure the subject and the predicate, just as they are called in grammar. Clearly, in the above examples the subjects are: *men, flowers, John, Peter,* and the pronouns; the predicates are the verbs: *eats, laughs, puts on weight, has a good time,* and other parts of the sentences that incorporate words complementing the subjects, such as *mortal* and *beautiful*. The subjects are designated by the lowercase letters w, x, y, and z, and these symbols are called arguments. The uppercase letters F, G, and H refer to predicates; they are the functional symbols that quantify. According to this scheme, the propositions are presented as follows:

| The flowers are beautiful | F x | Men are mortal | G z |
| Women are beautiful | F y | Monkeys are mortal | G w |

Where x, y, z, w are flowers, women, men, monkeys, and F and G mean "are beautiful" and "are mortal," respectively. For logical quantifiers, symbols are also used, such as:

For all x we use (x)

For some x we use (€ x)

So the propositions:

all the flowers are beautiful   would be   (x) Fx

some Americans are black   would be   (€x) H x

where H is the predicate "are black."

However, these propositions with quantifiers are molecular and imply two predicates each: the first one, to be flower (I) and to be beautiful (F); the second, to be American (J) and to be black (H). Therefore they should be expressed, respectively, as:

$$(x) (Ix > Fx),$$

$$(€y) (Jy \cdot Hy)$$

Some examples facilitate a comparison between the two logical methods: The proposition "if Peter laughs(L), then he has a good time(M)" is expressed in functional logic as L x > M x. In this case we would not be adding anything useful to propositional logic, because the given proposition is particular and it doesn't contain variables (x,y) or quantification.

If we analyze the so-often-cited syllogism shown below, we have:

Propositional Logic	Functional or quantifiable logic
p "All men are mortal"	(x) (N x > G x)
q "Socrates is a man"	N y
r "Socrates is mortal"	(p.q)>r ((x) (Nx>Gx) • Ny) > G(y)

where N is used for "to be man" and G for "to be mortal."

In this example, quantificational logic facilitates the proof, which is not the case for propositional logic. Although demonstrations in quantificational and propositional logic are of a similar nature, some rules should be added and the insertion rule must be modified It should be noted that quantificational logic is of much higher value for mathematical reasoning.

3-2 COMPREHENSION AND METALOGICS

Historical development of the foundations of mathematics has proven to be a formidable task, which involves subtle difficulties that have not been completely overcome.

We can refer to the development of mathematical logic, using the term "dimensions of logic," as done similarly within scientific disciplines. The great philosophers of logic, as Gottlob Frege and Bertrand Russell,[1] dreamed of creating a symbolism and a group of axioms and definitions that would be totally consistent and complete as a foundation of mathematics. Russell found contradictions in Frege's system and Wittgenstein[2] in turn did in kind with Russell and Whitehead's *Principia Mathematica,* which, after being perfected by its original authors, was finally considered consistent.

Nevertheless, the great mathematician Kurt Gödel, with his theorem of incompleteness,[3] proved that the limitation of incompleteness is inherent in logic and that there are mathematical problems that cannot be solved by a coherent set of procedures and given rules. A simple example of intuition and mathematical understanding, which can be useful as an illustration of the significance of logic's limitation as a symbolic mean, is presented by Russell in *The Art of Philosophizing and Other Essays.*[4] Russell refers to the existence of irrational numbers that were sensed intuitively by the Greeks for geometric reasons, when considering the reality of numbers like the square root of 2, which have a geometric expression by means of the Pythagorean theorem, as an "unknown number" whose square is equal to the area of the square of the hypotenuse of a right-angle triangle that has sides equal to 1. (See Figure 1.)

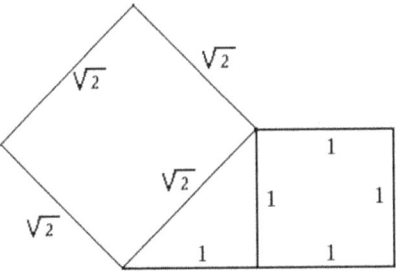

Figure 1: GEOMETRICAL EXPRESSION OF THE SQUARE ROOT OF 2.

$$(\sqrt{2})^2 = 1^2 + 1^2$$

[1] Bertrand Russell, *The Philosophy of Logical Atomism* (Open Court Publishing Co., 1996).
[2] Ludwig Wittgenstein, *Tractatus Logico Philosophicus* (Routledge Classics, 2001).
[3] Roger Penrose, *The Emperor's New Mind* (Oxford University Press, 1989).
[4] Bertrand Russell, *The Art of Philosophizing and Other Essays* (Totowa, NJ: Littlefield Adams Publishers, 1968)

Russell presents us with an arithmetical proof of the significance of these irrational numbers like the square root of 2 as follows:

> The Greeks soon discovered that there is no such a number. You can easily persuade yourself of this. The number cannot be an integer, because 1 is too small and 2 is too big. But if you multiply a fraction by itself, the answer will give us another fraction, not an integer; consequently there cannot be a fraction that multiplied by itself, gives 2 as an answer. In such a way, "the square root of 2" is neither an integer, nor a fraction. What is it then? The mystery remained; but mathematicians hopefully continued using it and speaking about it, trusting that someday its meaning will be discovered. Finally this expectation was justified.

We conclude that mathematical understanding in a global form goes beyond its symbolism; it is outside of the logic in a new dimension.

In a similar way in the world of theoretical physics, Albert Einstein proposes time as a new dimension that cannot be considered independently of space, like in Newtonian mechanics; rather, it remains hidden within our imagination, which doesn't conceive space in four dimensions. The analytical presentation of the theory of relativity originally restricts it to the framework of mathematical symbols of tensor algebra, revealing its characteristics by means of the Minkowsky representation, which reduces the space dimensions to two in a plane (or to one in a line) and the third perpendicular to the plane (or the line), as time. Each physical event is represented by a point of "space coordinate" that includes time, taken as space (c. t), c being the speed of light (see Figure 2).

We cannot imagine, no less draw, a space outside of the three dimensions in which we are immersed; however, it is conceivable to reduce ourselves to bi- or mono-dimensional beings, to explain the extra dimension as temporary. In modern times, after Minkowsky, we use the third space dimension to represent time, reducing our space with our imagination to two dimensions. This space model of so-called "space-time" facilitates understanding the theory of relativity.

In an inverse situation was medieval man, who, before geographical and astronomical discoveries, considered the surface of the Earth flat and the third dimension related to the future of our spirits, which would move after death to heaven or hell, according to the direction they would take.

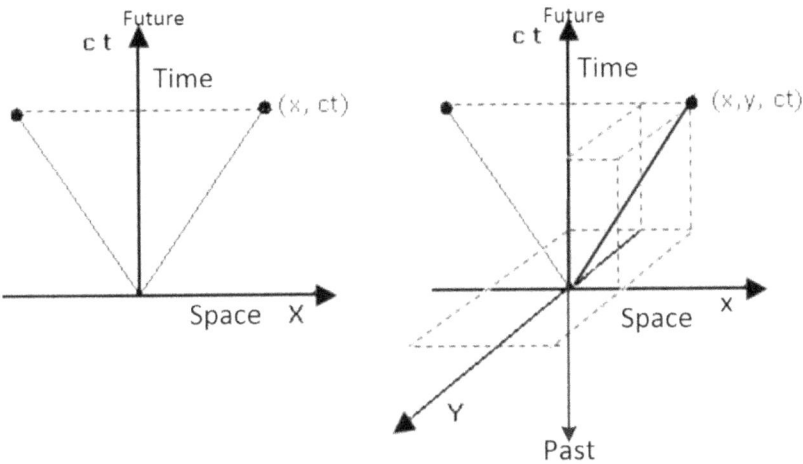

Figure 2: MINKOWSKY'S SPACE-TIME IN TWO AND THREE DIMENSIONS

Let us imagine that our logical ignorance is educated by going outside of its dimensions configured and reduced within the symbolism of language. The metalogic, as its prefix indicates, is located outside of itself, and is inexpressible. Human understanding extends itself in a dimension farther out than mathematical logic and its symbolism. On the other hand, the power of logic is like that of the clock, if one wants to use a mechanical simile, or as the modern electronic computer; it goes no further.

Comparisons between the artificial intelligence (AI) of computers and the human mind provide us with a vision in that dimension that we would like to explore. The extraordinary advances of AI as of late refer to memory capacity and processing speed. At issue is whether biological learning can be replicated by computers. Also, one questions if the computer, which works in a sequential form, can acquire some mental abilities, as it is designed with parallel circuits which are interrelated.

An illustration of the lack of understanding of modern computers is presented in the final part of a chess game that Roger Penrose[1] transcribes in his book *Shadows of the Mind*. At the final game position shown in Figure 3, black has a material advantage of pieces that seemingly gives it a great advantage, while the white pawn's position is strategically favorable.

A powerful computer that has defeated masters, playing in this example with the white pieces, makes the wrong move and takes a black tower with

---

[1] Roger Penrose, *Shadows of the Mind* (Oxford University Press, 1994).

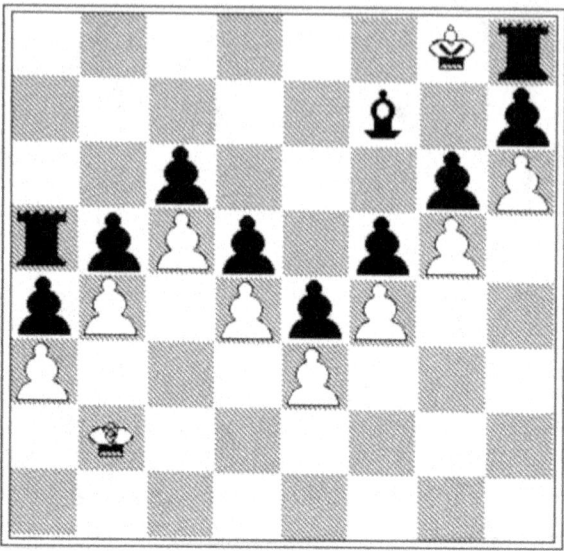

Figure 3: CHESS AND UNDERSTANDING

a pawn, sacrificing the overall strategic position of its pawns. On the other hand, an experienced player intuitively senses the strategic importance of his position and keeps moving the king, tying the game. Undoubtedly human understanding in general and mathematical understanding in particular cannot be replaced by computation.

Frequently, we use computational or mechanical procedures when we need to obtain immediate results. In these cases, we don't employ the power of the mind; we act as machines, so we can look very intelligent, without in fact understanding anything. Strictly speaking, innovation is not obtained by using computational or deductive rules by means of an algorithm already established. Creativity has a different feature than computation; it is a consequence of understanding, which so far has not been achieved by artificial intelligence.

3-3 POST-PHILOSOPHY AND LINGUISTICS

To return to symbolism, in search of other internal dimensions, we should return to a position of self-consciousness, to examine the formation of fundamental concepts such as time. Psychologists such as Jean Piaget[1] preferred to take a behaviorist position, that is to say, from the point of view of the examination of external or public phenomena, to

---

[1] Jean Piaget, *El desarrollo de la noción del tiempo en el niño* (Mexico: Fondo Cult. Ec., 1992).

describe the emergence of basic concepts such as time in the child. Introspection is mainly an adult characteristic; on the other hand, experimental and theoretical psychology uses in general a mixed methodology, considering external or public observation as well as private or internal data. The attitude of behaviorism, which pretends to be more "scientific" and thus "objective," results from the desire to transfer in absolute form the methods of the natural sciences to the analysis of the mind.

In any study that we develop on the formation of mental concepts, we have to begin from internal events or phenomena that have a strict private value that is absolutely individual. Distinguished philosophers, such as Russell, have used the sensation of color as an example of a primitive element with a private character in the formation of language. Recently Nicholas Humphrey,[1] in his book *A History of the Mind*, has explored interesting psychological aspects relative to the ideal experiment of "inverting colors" analyzed by Locke and Wittgenstein. An important aspect that is not usually emphasized refers to the consequences of the strictly private character of these primary elements of knowledge, such as the sensation of color, that are on the fringe of language.

Color may be analyzed from a physical point of view, using the wave theory of light, in a way that we don't doubt that color is completely defined, even numerically, by the wavelength of light.[2] Psychologically, this definition is not as transparent as it first may seem. The sensation of color—for example, blue—is a private fact and therefore not shared among different individuals. The word *blue* is the symbol that we use to come to an agreement on how to describe a fact, but this agreement is only formal, because there is no way to know if the sensation that I learned to call blue is the same one that others have when they say "this is blue." We only come to an agreement on the symbol for "this." So we pass from a private fact to a name, a symbol that is public by an ostensive, shown definition. This doesn't guarantee us that our sensation of blue is the same one experienced by all. Symbols such as *red* and *blue* have public validity, but that doesn't necessarily imply that they also have private, individual validity. This is the limit of language.

It is worth observing that this analysis of color sensation has nothing to do with physiologic or medical studies on defects of vision, such as the so-called Daltonism, as there are people who don't detect colors, but rather only differentiate bigger or smaller intensities of light—that is to

---

[1] Nicholas Humphrey, *A History of the Mind: Evolution and the Birth of Consciousness* (New York: Simon & Schuster, 1992).
[2] Richard P. Feynman et al., *Lectures on Physics, Vol I* (Boston: Addison Wesley Longman, 1970).

say, they only see in black and white. This is a publicly observable fact. If a child is shown the colors of the rainbow and taught the words corresponding to these colors—for example, in the order of the spectrum, red, yellow, green, blue, violet—even though his sensations to the colors may not correspond to those of his teacher, he will express all colors correctly. It could happen that the boy, for example, doesn't distinguish red from blue, which is to say he sees them alike; in that case this defect would be detectable or public. It is important to know that differentiating two sensations, although they are private in nature, acquires a public character in language and in logic. Diversity is the origin of thought.

Language and its modalities symbolize the diversity and the limit of that game; it is within our capabilities to differentiate events however subtle they may be. Wittgenstein, in his *Philosophical Investigations* and *The Blue and Brown Books*,[1] in a nonsystematic fashion abandons the logical consistency of his previous *Tractatus* and gives place to a concept of language that expresses human happenstances that cannot be reduced to the rigidity of mathematical logic. I don't consider that this second view from Wittgenstein opposes the first one; just that with the acquired authority on the subject matter from his famous *Tractatus*, which impacted the very foundations of Russell's *Principia Mathematica*, he earned the luxury of playing devil's advocate to his own ideas and to those of classical philosophy, discovering aspects of language that reflect the complexity of human relationships that go a lot further than its symbolic structure.

Contemporary philosophy has embarked on that story and sailed away with it. In a particular way this philosophical trend is similar to the artistic evolution that, looking for novelty, has produced concepts in the plastic arts that move away from the description of reality, returning to primitivism or to impressionistic themes that suggest indefinite psychic moods. Manuel Cruz, in *Contemporary Philosophy*, describes brilliantly how philosophy returns at this last stage to its literary origin. As the canteen of rationalism is emptied, new philosophers look for liberation from logic and especially from science, which has reduced their habitual sphere. Russell's black cat is trapped, and only if we let it loose can we look for it again (see Section 4-2).

Among other ideas, it has been proposed that philosophy disappear as a specialized discipline and return as part of the general field of knowledge. This strategy could be implemented in the large universities, where the philosophy departments would blend into the other different disciplines,

---

[1] Ludwig Wittgenstein, *Philosophical Investigations: Critical Essays* (Blackwell Publishing, 2001) and Ludwig Wittgenstein, *The Blue and Brown Books* (New York: Harper & Row Publishers, 1958).

leading to greater cultural plurality that would mutually enrich the specialized disciplines and the philosophical concept of knowledge in general.

Dilettantist philosophy in the second half of the twentieth century is a consequence of disillusion produced by the impossibility of finding a perfect logical language that replaces ordinary language, with all its modalities and interpretations that in fact constitute its biggest wealth—a product of the evolutionary mechanism to which it has given rise. Richard Rorty[1] reminds us of the pictographic origin of language, in the idea that "the activity of uttering sentences is one of the things that people do in order to cope with their environment"; it is a necessity, there is no possible exile from language, this is the price we pay to be social. However, we should recognize that the origin of knowledge is individual. It resides in originality, and to transcend and therefore have relevance, it becomes public.

Language reflects the nuances of the mind between its two most characterized aspects, thought and feelings, and gives validity to literature and arts on an equal footing to science. Philosophy should conserve its universal value that has been its historical profile, recognizing its pluralism without destroying any of the imprints of language. Thus, post-philosophy will be a perennial philosophy, a catalyst for knowledge. It will be neither new nor old, without delirious pretenses, but not reduced to an irrelevant word game that produces confusion and mental chaos. Literature has always been near to philosophy and now it is back within its realms. "Existentialism" can be wordy for the "rationalist" or for the "man on the street," but for the "bohemian" it is poetry and for its authors it is philosophy. Mystics prefer silence.

3-4 SCIENTIFIC CONCEPTS, PHYSICAL AND MENTAL EVENTS

Moving on from elementary sensations such as color to others of a greater complexity, and taking an introspective attitude, let us reflect on the order of memory, in connection with the formation of the concept of time. We observe that this, at least from the psychological point of view, involves a group of memories, or, plurally, groups of memories, with which we denominate the past. What is the base that sustains these events or memories and allows their classification? From the physical point of view, we believe that memory forms by means of circuits of neuronal communication. Nevertheless, biological memory, unlike the memory of an electronic computer, is far from being a printed symbol-

---

[1] M. Cruz, *Filosofía contemporánea* (Taurus Ed., 2002).

ism, which has a given spatial arrangement, in a sequential order. Systemic theory of communication networks, based mathematically on the theory of "chaos,"[1] proposes by means of nonlinear functions a series of communication states that dynamically follow behavior patterns, which vary slightly around a center or "attractor" that characterizes them, passing abruptly to a new "basin" that has different patterns. We shall have occasion later on to present these theories in some detail.

In the environ of these ideas, the ordering of events within the memory in linear or serial form that characterizes the concept of time doesn't seem to be related with internal physical states that can be ordered spatially. More feasibly it can originate from a communication pattern within a network of punctual memories, which defines a global serial trend. This direction is a consequence of the action of external events of diverse origin that repeat themselves in the same sequential order, giving place to patterns of communication that we call mental habits.

Our imagination brings us easily to discover that some events that happen invariably in nature, such as sunrise and sunset, the seasons, the falling bodies, and the growth of plants and animals, constitute the external stimuli that give origin to the establishment of psychological time and its arrow. This concept is slowly established as these groups of events are settled in parallel, providing a multidimensional order inside a psychic space, characterized by communication patterns.

Undoubtedly this mechanism is also common to other animal species that share with us an intuitive concept of time. Mankind advances in this temporal order by means of symbolism, which plays a crucial role in mental relationships. Symbols give to the mind internal serial benchmarks that facilitate the ordering of memory, providing time with an arrow or direction. For that reason, the concept of time in a child is developed slowly and is tied to the learning of language.

The elements of memory, at an early age, are dispersed; language in general and numeric symbols provide references that are increasingly firm guidelines that establish sequential memory. This process continues from the individual and social point of view, giving place to wider concepts of time, such as age and the calendar, as well as time in science, which only refers to the external world and its order as expressed in the laws of physics.

To understand scientific knowledge that relates the facts of the world, it is not enough to employ deductive logical inference that is fundamentally symbolic; it is necessary to use inductive logic that looks toward the future by means of the analysis of the past.

---

[1] I. Prigogine, *Las leyes del caos* (Editorial Crítica, 1999).

FOUNDATIONS OF KNOWLEDGE

4-1 CAUSE AND EFFECT

When we speak of cause and effect, we are not referring to deductive logic that relates propositions, but to relationships between facts. What is the true reach of these two words, which seem to govern our lives? Is it to facilitate our existential plight, or is it to submit us to their relentless laws?

Their origin goes back to the dawn of human society and is captured in the mythical theories of primitive societies. What is the relationship between mythology and science? At first glance, there is none; it seems that they are opposed to and exclude each other. However, their dialectic leads to the synthesis of knowledge. They both have a purpose: the dominion and understanding of nature. They look toward the future for a practical objective, and they are based on past experiences, to devise myths or theories that facilitate mental order.

We usually say that primitive societies based their organization on superstition, a word that, according to the Spanish dictionary, means: "Belief in strange things to faith, and contrary to reason," which says in a different way that superstition is opposed to established mythology, faith, and to the rules of deductive logic, or reason. Using modern terminology, we would say that superstition follows neither the current paradigm in a society nor mathematical logic. To what extent, through the ages, does humanity still remain captive, enslaved by its own mythology?

Paradoxically, Western science has much more affinity with Eastern philosophical mythology, such as Hinduism and Buddhism, than with the Judeo-Christian myths that propitiated its blossoming. Greek thought, which was fundamentally logical, lacked in its mythology the belief in an underlying order in nature that had not yet been discovered. Its gods didn't direct the world, nor did they create it; they were a voluptuous court that enjoyed a paradise that symbolized human desires. On the other hand, the God from the Bible was wise and almighty, and His creations should follow His unchangeable laws.

In Eastern mythology, there is no personal god whose nature transcends the universe. Nobody governs the world; its order is found in the dynamics of its own cyclical evolution.

Time in the East is static and dynamic; its cycles imply the eternity of the universe and the unimportance of individuality. History is not defined with a beginning and end, its calendar has an unlimited extension, dates are not even registered; nothing is new or old on Earth. In that universe, causation is limited to our own individuality that is apparent; it doesn't correspond to reality that is unique and universal.

In Eastern philosophy, there is no room for determinism of the laws that supposedly govern the universe, which is an organism that evolves in its diversity between emptiness and existence, life and death, the concept of self and universal unity, time and eternity. Western science, which pretends to find universal order, and the eternal laws that govern it could not have come from an Eastern mythology that neither denies nor affirms them, nor considers them important, because its objective is not to dominate nature, but rather to gain understanding through individual transcendence, by meditation and suppression of desire, the source of suffering. Paradoxically, Western science comes closer in its later developments, especially with quantum theory and its uncertainty at a microscopic level, to a world less deterministic and more in tune with Eastern mythology.

On the other hand, Greek philosophers expressed their admiration for abstract reasoning in geometry and the mathematics of the Pythagorean followers, as well as for the reality and beauty of the Platonic ideas. There was no room for the prosaic world of facts, which were believed, by their own imperfection, to confirm the absolute truth of ideology. Western science rejected this Platonic belief and looked for the laws of nature based on facts, taking from the Greeks only their mathematical logic to explain them.

Christian mythology entered into conflict with modern science even though it had unwittingly propitiated it. Modern science, after five centuries of development and improvement, returns to the Platonic world, becoming more abstract and closer to mathematics.

Contemporary physics, in its interpretation of the world within the confines of space-time, gets further every time from anthropic concepts, and in its eagerness to find a unified theory of its two major branches, quantum theory and general relativity, comes increasingly closer to an abstract, purely logical formulism, based in mathematical entities like multidimensional "strings" and supergravity in ten or eleven dimensions. Stephen Hawking,[1] in his scientific writings, emphasizes his "positivist" position that can be interpreted by its context as the justification for mathematical hypotheses of science, regardless of how far away they may be from a possible interpretation at a human scale, as long as logically they are compatible with the facts of the world.

Strictly, this positivist view has been the base for natural sciences, especially physics. However, this position can lead us to a "numerology," or mathematical games that are so far from our psychological world that we lose all contact with reality to a point that, in spite of achieving a unified theory, which may or may not be favored by new empirical evidence, there will be a diversity of theoretical possibilities that, by lacking meaning in themselves, will be equally eligible.

Not all science is mathematics, and mathematics itself is not knowledge. After the formulation of Newtonian mechanics, by means of differential calculus, big theoretical developments have been carried out inside the framework of classical physics. In parallel, chemical science, based on observations of the diversity of the natural world, found the conceptual bases for the structure of matter within an intuitive frame of mind that contributed fundamentally to the subsequent formulation of the mathematical theory of atomic physics.

In the late twentieth century, new experimental and theoretical developments within chemistry, biology, and the social sciences have opened new horizons that unfold into a science where indetermination of the phenomena in complex systems is studied in a global form, called systemic science—that is to say, considered collectively and not causally, starting from elementary facts that supposedly determine the total behavior.

Everything seems to indicate that the traditional scientific attitude, called reductionist, which uses deterministic causal chains that strictly are only applicable to idealistic simple systems, has arrived at an evolutionary threshold, in which new systemic developments impose an uncertainty that goes further than the current formulation of quantum mechanics and statistical thermodynamics of equilibrium.

On the other hand, from a social and political point of view, the results of Western science, despite all its undeniable achievements, have brought

---

[1] Stephen W. Hawking, *The Theory of Everything: The Origin and Fate of the Universe* (New Millennium Press, 1996).

humanity to its current situation of environmental destruction, a direct result of the population's exponential growth, and of an economy that contributes to waste and doesn't satisfy man's fundamental necessities. This reality takes more and more a collective consciousness, and eventually may be able to promote socially a worldwide synthesis of Eastern philosophical ethics and the scientific paradigm that moderates the materialistic desire to dominate nature, which imposed the deterministic laws.

4-2 DEDUCTION AND INDUCTION

All of these historical concepts regarding the development of human knowledge, and in particular the scientific aspect of it, bring us to define two mechanisms that are usually called deductive logic and inductive logic. These processes don't refer to different directions but rather complement each other. Induction clearly implies employment of deduction; in deductive logic, induction is not even mentioned. For the sake of clarity and transparency in language, it is best not to refer to the induction process as properly logical and to reserve this term for deduction.

Induction explicitly refers to facts; on the other hand, logic is a symbolic procedure that allows us to establish deductive rules that relate the true characters of propositions. Logic in itself doesn't provide knowledge; it is outside of reality. To establish knowledge, we must appeal to observation of facts and their definition by means of propositions, which we call true when they correspond to the facts and false when they do not. The group of truthful propositions constitutes knowledge that we call scientific when it is obtained by means of the induction process and relates propositions in a coherent form, using logic.

Generally, induction is considered restricted to an observation process that relates facts, predicting its repetition in a temporal order with a certain probability. On the other hand, the imaginative process that seeks to explain this relationship is denominated intuition; it constitutes a mental element that is close to the individual and the collective unconscious thought process that is manifested suddenly, as in the case of Archimedes, when he exclaimed "Eureka!" and discovered the flotation principle. It seems convenient to include this intuitive process as a characteristic of the induction process, by means of which the fundamental premises or scientific principles are proposed as hypothesis, from which the propositions of science may or may not be deduced by logic. This methodology, which we would call "trial and error," allows us to select the principles logically compatible with experimental data. Moreover, these groups of principles established by the process of induction constitute the scientific theories that will be favored or not, with predictions of new experimental facts found in the development of science. In this sense, induction looks toward the future as an extrapolation of the past.

Historical consolidation of the inductive process definitively marked the extraordinary modern blooming in the acquisition of knowledge. Frequently, reference is made to this scientific methodology, as a principle that is usually called objectivity, which allowed the transition from Greek philosophy, based mainly on logic, to the Renaissance philosophy that gave place to modern science. For the Greeks, logical truths were eternal, unshakable; if the facts didn't agree with their conclusions, this would only show nature's imperfections.

The mathematical-logic structure of Euclidian geometry is undoubtedly imprinted in the scientific thought process. Also, its conclusions on the properties of space were useful for proposing the laws of movement in classical mechanics. However, in geometry, as its name indicates—measurement of the Earth—there is an objective element implied. That is to say, there is a premise in which its validity is inductive; it is based on observations of the properties of space that are expressed numerically in measurements.

This empiric aspect of geometry was not explicitly understood by the pioneers of the new physical science, classical mechanics, and probably much less by the Greek geometry scholars who considered their reasoning free of the imperfections of the facts of the world. The invention of new geometries, based on different postulates and the confirmation of the theory of relativity, using non-Euclidian geometry invented by Bernhard Riemann, transferred this discipline, traditionally considered strictly mathematical, to the physical science domain.

It is worth wondering whether all of human knowledge, including logical principles, has its origin in observations of the physical, or external, world, or whether it is permissible to think of its intrinsic validity, regardless of experimental confirmation. Undoubtedly, arithmetic propositions such as $2+2=4$ have a different verifiable value than those such as "the Earth rotates around the sun." However, the Pythagorean theorem, which expresses a relationship of measures of the sides of the right triangle,

$$a^2+b^2 = 1^2$$

is deducible logically from the premises of Euclidian geometry, and was "discovered" as a property of space and used by other cultures previous to the Greek, such as by the Egyptians.

In modern times, it has been found experimentally that Euclidian geometry has a restricted validity limited to our habitual world that corresponds to an approximation of a more general theory, Einstein's restricted relativity theory, which proposes a space of four dimensions. Space-time implies the interrelation of four dimensions to define an event, in the same way that the usual three dimensions of space define a position. Again a theorem of Pythagoras appears, in four dimensions,

$$a^2+b^2+d^2-(ct)^2 = l^2$$

where a, b, d are the space dimensions and (ct) represents the time dimension, multiplied by the speed of the light c that reduces it to a space dimension, and where l represents an interval between two events in the so-called Minkowsky space. At a subsequent stage, Einstein presented his general theory of relativity, which reduces gravitation that implies acceleration, or change of speed, to a property of space-time, following the Riemann geometry proposed one century before relativity. This is how geometry passed in a definitive manner to the physical sciences.

Traditionally, philosophy included all of human knowledge; however, science has evolved its domain and dimensions to such an extent that philosophy is now considered as a dynamic for thought rather than knowledge. It could be interpreted that this trend of science, which invades places previously thought to be the absolute domain of pure reason, confirms that the whole of human thought is based on facts, and that there are no innate or preestablished ideas.

Bertrand Russell tells of philosophers looking for a black cat in a dark room, while theologians do the same in a room where there is no cat. It may be added, that the scientists who had a flashlight found the cat, while the others continued looking for it. Hurrah for the philosophers and the theologians who won't finish the game, because that is what it is all about.

It remains to be seen if the theologians are really where the cat is not. One might think that they are dealing with another dimension outside of physics and reason. Unfortunately, this dimension is related traditionally with human behavior, with human desires that darken its meaning. Einstein, of Jewish tradition, defending his theory of relativity that is a logical edifice, used to say by way of criticism, referring to the indetermination of quantum physics: "God doesn't play with dice" (see Section 5-6-2 A). Einstein's God is the personal God of the Bible; he is a super-being, only He can fix absolutely the initial conditions of a system, making it possible to determine the future or the past. Mankind cannot do it, so we have invented time and its arrow.

Concerning cosmological models that have emerged from relativity, Einstein was initially inclined toward a stationary model of the universe, by introducing the cosmological constant in his equations, with its repellent effect balancing the attraction of matter. Subsequently, the astronomical discovery of the distancing of galaxies suggested the expansion of the universe, and at the same time Alexander Friedmann and Georges Lemaître found other solutions for the relativity equations that agreed with that model.

Ilya Prigogine[1] relates an anecdote from his friend Lemaître, that on a certain occasion, when he was referring to the initial state of the expanding universe, Einstein replied to him: "That reminds me too much of the Genesis; I can tell you are a priest." It is well known that the big bang model, which Lemaître called the primitive atom, has been favored by new experimental facts, such as the detection of cosmic microwave background that comes with equal characteristics in all directions.

Penrose and Hawking[2] consider that relativity's mathematical model demands that time have a beginning, which agrees with Saint Augustine's Christian thought that says "that before God made heaven and earth, He didn't make absolutely anything." On the other hand, many physicists don't agree with the idea that time has a beginning and an end; there is no lack of other theories that end up resonating with them. For example, models of multiple dimensions other than the relativistic space-time dimensions, which expand to superastronomic dimensions, give place to a superuniverse, where universes like ours appear and die like bubbles inside a "liquid phase" that represents a vacuum.

4-3 SCIENCE AND SUPERSTITION

Prejudices of any kind have never been a good guide for acquiring knowledge, and not even the big geniuses are exempt. Just as religious prejudices have been a great historical obstacle for progress, so today it is their opposite, which may be called materialism, that poses the problem. Science itself, like any ideological system, can constitute a barrier to gaining new knowledge. Let's not forget that science has its origins in mythology and superstition. Physics originated from astronomy and mythology of the personal God, chemistry from alchemy and magic, and medicine from sorcery or witchcraft. It is worth asking ourselves:

---

[1] Ilya Prigogine, ¿Tan sólo una ilusión? Una exploración del caos al orden (Tusquets editores, 1997).
[2] Stephen W. Hawking et al., *The Future of Spacetime* (New York: W.W. Norton & Co., 2002).

Are the causal relationships among facts in the field of superstition or of knowledge?

Many concepts that seem alike vanish if we intend to define them. Saint Augustine's story about time is famous: "If nobody asks me, I know it; if I want to explain it to somebody who asks me, I do not know." Something similar happens to us with the relationship of causality. The concepts of time and causation are bound in indivisible form: Cause precedes effect, it is a necessary condition to establish the relationship, but it is not enough. Induction comes from the repetition of facts in a determined order and as such it only indicates a probability.

On the other hand, as David Hume[1] states, causality is different from the relations of time and space, as they are inferred directly from our senses, and on the contrary, we never perceive directly causal relationships. When we say A causes B, we simply affirm the observation that two phenomena, A and B, successively happen; our inclination to believe that they are necessarily tied to each other is a consequence of observing their repetition, which forms habits of association of memories or establishes certain patterns of mental communication. As Hume states, "the supposition that the future resembles the past, is not founded on arguments of any kind, but rather it derives entirely from habit."

We can then infer that the law of causality, A causes B, is simply a law of probability: If two phenomena are repeated a sufficiently high number of times, we conclude, with a calculable probability, that we can expect this to happen again. These observations tell us nothing about the intrinsic relationship between the two phenomena. Strictly, it is not possible to define that connection among the facts. Only by means of the inductive process, we propose a logical order between a group of axioms or postulates, from which other propositions are deducted and confirmed by the facts. This scientific method, in common language, is expressed by saying that we have an explanation of the facts; otherwise we consider our hypotheses to be false.

Causation, in its primitive form, is expressed by the superstition that evolves historically, by means of symbolism in mythology, proposing an explanation of the world whose elaboration grade is manifested in the multiple dimensions of philosophy and science. Does the original conception of superstition in the causal relationship subsist within elaborate modern theories? It seems that scientific explanations only transfer a simple causal relationship among facts to a chain of causal elements

---

[1] Bertrand Russell, *The History of Western Philosophy* (New York: Simon & Schuster, 1945).

that increase their credibility. This philosophical problem is part of the dynamics of thought, and its only solution, as Wittgenstein states in his *Tractatus,* is not to solve it:

> There are natural laws. But of course, such a thing cannot be said; it is shown.

> We cannot compare any process with the course of time; this doesn't exist, but only with another process.

> The process of induction is the process of assuming the simplest law that can be made to harmonize with our experience.

It seems that all these arguments can lead us to total skepticism. Some examples, ordered from a smaller to a larger degree of complexity, will give us a more intuitive perspective on these basic concepts of induction—a process whose validity is beyond logic and belongs to a dimension without which science is impossible.

Natural phenomena were interpreted primitively in a simple causal form. These beliefs still subsist in rural communities that are directly in contact with nature and that, because of their cultural isolation, still maintain certain mythical characteristics. These beliefs can also be found in "civilized" societies, where "magic powers" survive, especially in connection with drugs and alternative treatments to cure illnesses.

The influence of celestial bodies on our destiny, the reading of the future by means of palmistry, and an infinite number of other, similar beliefs—not to mention political and religious myths—are common in today's world. Among these, I find it illustrative to analyze the belief in the influence of the phases of the moon on the development of plants and on the effectiveness of agricultural techniques. Many peasant farmers who inhabit the tropical areas of America associate the waxing and waning phases of the moon with plant growth, which is favored or not according to its luminous cycle. They conclude that it is not favorable to sow or harvest in a growing moon season, if one desires good and abundant fruits.

The aspect that interests us in our analysis is the causal relationship between the variation of the brightness of the moon and the growth of the plants, when assuming the coincidence of the two phenomena. It is considered that plants, when diminishing their growth in the waning of the moon, give place to a harvest that grows and matures better.

Scientists can demonstrate that this relationship of causation is without foundation, measuring all kinds of parameters that do not corroborate it. If they are physicists, they will say that this relationship is not deducible from the gravitational influence of the moon, according to the relative position of the three celestial bodies, during its revolution around the Earth, and that the lunar gravitational effect on the terrestrial surface is manifested mainly in the tides, which have a cycle of twelve hours, according to the speed of rotation of the Earth on itself.

However, for the peasant farmer this causal relationship will continue being evident. It is simply fascinating, it is imposed, and it has a mythical character. This example shows schematically how the causal relationship in itself is assumed and deceiving; only the induction process can have credibility in an environment of logical parameters that in this example would be in the realm of natural sciences.

The most diffused myths of all times, from the tribal cultures through modern societies, are related to human health. What are the causes of illness? This is a question as remote as the cultural emergence of the human mind. In the most primitive tribal communities, sorcery is a basic cultural element, from the pragmatic point of view, as well as a means to constitute hierarchies within their organization. The history of medicine shows the relationship between the testing of physical means, such as the employment of natural drugs and corporal treatments, and the use of magic procedures, justified in beliefs or myths. These beliefs serve multiple purposes, functioning as an explanation or theory of the world, as a practical element that allows influencing the course of nature, and, perhaps more importantly, as a means for social cohesion.

In the development of medical practices is evidenced the concept of cause as the basic element in the search for a cure. Modern Western medicine followed this pragmatic method, developing its foundations in the induction process, which has obtained so many successes in the new physical science. The adoption of this new attitude, based on the principle of objectivity, was slowly banishing magical or superstitious procedures that had persisted through the ages. Just as the physics of Galileo and Newton arose in large part from knowledge gained by the employment of the telescope, medical science took a new direction with the invention of the microscope, which allowed the discovery of microorganisms. The search for the causes of some illnesses in the microscopic world undoubtedly provided the key that made possible Western medicine. With the famous studies of Louis Pasteur, medieval times were left behind, such as the black plague that decimated Europe's population and took it to a near point of disappearance. The "cholera" or "wrath of God" left its name to that plague that justified the rituals of the flagellants seeking victims among Jewish communities or "witches," to appease divine punishment, opening the way to a new era that, little by little, has made disappear the inherited mythology of the Greco-Roman tradition, supplanted by the Judeo-Christian beliefs and superstitions of the primitive Northern barbarians.

Western medicine, inspired by logic and observation, derived in the new millennium, like all science, a recognition of indeterminacy of natural events, to a greater or lesser extent away from the causal chains that are idealized in classic science, especially physics. As more complex systems are considered, such as in chemistry and biology, and certainly in psychology, it is proved that causation is not the supreme law of nature. The experienced physician knows very well the reaches of the idealized knowledge taught in the university, and values the indeterminate factors that intervene in the health of patients. Medicine in general hopefully evolves toward less rigid concepts than those imposed on it by dehumanizing modern technology; traditional medicines, such as the practice of Chinese acupuncture or the Ayurveda from India, deserve increasing respect and study.

The Hindu doctor Deepak Chopra,[1] who has a dual knowledge of the Ayurveda and Western medicine, has popularized the integration of the two cultures, emphasizing the importance of considering the union of mind and body, and proposing medicine as a way of life, and not simply as a system for the prevention and cure of illness, based on the fear of death and pain.

An understanding of global or systemic concepts is complementary and not opposed to the analytic method, which is sometimes pejoratively called reductionary. Undoubtedly its methods are complex, as are the systems it seeks to understand; it doesn't expect to manage the future, its wisdom is one of expectation, not of certainty. Are we then returning to superstition and witchcraft? Obviously not; the recognition of the limitations of strictly causal methods is simply admitting reality in all its complexity—a milestone on the road to knowledge that looks for new orders of greater hierarchy in nature.

The complex world of psychology and psychiatry, with the advent of Sigmund Freud and Carl Jung's psychoanalysis, had a great social impact that was evidenced in the popularity of concepts such as the uncon-

---

[1] Deepak Chopra, *Ageless Body, Timeless Mind* (London: Rider & Co., 2003).

scious and the subconscious and their importance in human behavior and education. The subsequent disillusion produced by the psychoanalytical methods, while practicing psychiatry, is simply a consequence of the fact that mental phenomena cannot be reduced to causal simplicity. As a reaction, psychiatry has passed to a physical-therapeutic approach, using psychoactive drugs with doubtful results in the individual patient. However, collectively it has been possible to successfully replace the degrading methods traditionally used in the centers of mental health.

It is interesting, the incursion of philosophy as a psychiatric method that tries to use "the useless knowledge"—philosophical illustration and its dynamics—not as an external method to its "patient," but by incorporating in the patient's intellect a wider, less personalized concept of our project of life. This new philosophical projection, presented by Lou Marinoff,[12] in his work *Plato, Not Prozac*, contrasts the traditional self-help books that cram the shelves of bookstores, which basically are limited to advice to their readers. Nobody can be taught how to think; we all must do it for ourselves—this is the essence of philosophy; it is its message. In this sense, philosophy can be a means to induce a healthy mind.

The great difference between artificial intelligence and the human mind is the same one that exists between a parrot and a philosopher, when applying a formula and understanding its meaning or, better yet, inventing it. There is no test that can prove that somebody understands, simply because understanding is a fact of private domain and a test is public. The impact of modern society on the mind is alienating us; it deprives mankind of its most precious gift, its understanding. It turns us into an object within the masses and transforms us into mechanical elements. Philosophy, on the contrary, is the dynamics of thought; it is a game that makes us human.

In the world of natural sciences, the meaning and scope of the systemic or global approach can be defined specifically, in contrast with the traditional reductionist method of causal chains.

---

[1] Lou Marinoff, *Plato, Not Prozac: Applying Eternal Wisdom to Everyday Problems* (New York: HarperCollins Publishers, 1999).
[2] Lou Marinoff, *Preguntale a Platon* (Barcelona: Edic.B.S.A., 2006).

PART *Two*

THE PHYSICAL OR EXTERNAL WORLD

5

SCIENCE AND INTUITION

All scientific theories, especially in physics, have two fundamental aspects that make possible their conception and development: their qualitative aspect is based on intuition and the quantitative one is expressed by means of mathematical logic. These are in turn the fundamental elements of the induction process: Intuition originates from experimental facts and from some facets of our thought process that by their own nature cannot be generalized, and, so to speak, float in the cultural environment, crystallizing in the creative minds of those we call genius. Mathematical theories are part of the cultural atmosphere preceding scientific gestation, and also evolve simultaneously in some cases. We consider that mathematical tools are indispensable in the formulation of a new scientific theory, even though they are not enough.

In the presentation of a scientific doctrine, an essentially deductive method may be used, just as in geometry, based on postulates and deducing theorems and conclusions that may affirm or refute its validity in connection with the experimental data. In a didactic way, one can emphasize the ideas and facts that made possible their enunciation: intuition and observation. Global understanding requires both focuses; however, the rigorous form is by nature cold and leaves us unsatisfied, because it doesn't illustrate to us how it could have arisen. In consequence, even though we appreciate its logical beauty, we don't feel the conceptual bases that are related to our own human nature.

In what follows, some fundamental scientific theories are presented; it is desired to highlight intuitive aspects of their historical development, making reference to mathematical methods with less detail, limiting their presentation to the essential parts. Some mathematical topics are developed in a more detailed fashion but are not considered indispensable for the continuity of the text; they are presented as supplemental, indicating their extension clearly.

5-1 CELESTIAL MECHANICS

These two words—celestial mechanics—have had a great impact on me since my childhood; they contain a contrast within our habitual world that is at our reach—the Earth and the outer world, the sky. They divide space and time into dimensions of our earthly existence and the spiritual dimension of our mind, transporting us to the confines of the universe that our senses notice only upon viewing a spectacular starry night, or with the meditation that transcends our individuality.

Modern life has denied these extraordinary impressions to most of the world's population that remains in the cavern of big cities. We have deprived our children of that extraordinary and fascinating observation of the universe that made it possible for humanity to attain its greatest logical prowess, mathematical physics, and its highest spiritual conception, mystic transcendence.

Mechanics, the foundation of mathematical physics, had its origin in the conjunction between the dynamics of the bodies in our immediate environment and the astral movements; it was the great synthesis that related the experiences of the falling of bodies—the apple in Newton's garden and the Galilean ring from the Tower of Pisa—with the observations of Nicolaus Copernicus and Johannes Kepler of the celestial displacements in the rutilant nights.

It is worth asking: What god aided these great creators of "celestial mechanics" that had avoided the major civilizations of the past? Those who in their observations had devised the calendar, relating the cycles of nature with the celestial movements in the sky that were explained by means of an egocentric cosmic model whose proud conception placed mankind in a privileged place, forgetting that it is but only a part of the cosmos.

In the Renaissance of Medieval Europe, very diverse cultural aspects converged. Perhaps the most important, in connection with science, was the intuitive certainty that the divine laws that were to be discovered and that should govern the whole universe, would only be deciphered by means of experimentation guided by mathematical methods. Greek geometry, the new developments of algebra, and the Cartesian representation of coordinates made possible a synthesis between arithmetic and geometry that projected time in spatial measures, giving place to the kinematics of bodies.

It is worth reflecting on the election of the coordinates that Galileo used in his famous experiments. Obviously in the free fall of bodies, one coordinate is the distance (e) traveled by that body. It was not so easy to designate the other coordinate that had to be related with space. Let us remember that in the sixteenth century mechanical clocks didn't exist and, on the other hand, Aristotle, the unquestionable authority of the time, affirmed that bodies fell with proportional speeds to their weights. There was no numerical concept for speed and its variation, or for time. Galileo's brilliance or intuition, whichever, led him to choose as a second coordinate natural numbers that designated intervals among events, which he assumed repeated themselves periodically, like his heart pulse or perhaps the oscillation of a lamp that opposed the inclination of the tower. The Cartesian representation revealed the relationship needed for space-time. In Figure 4, graph (a) shows us a curve called a parable, while in (b) we observe a straight line that illustrates the relationship e = k t^2, where k is a constant.

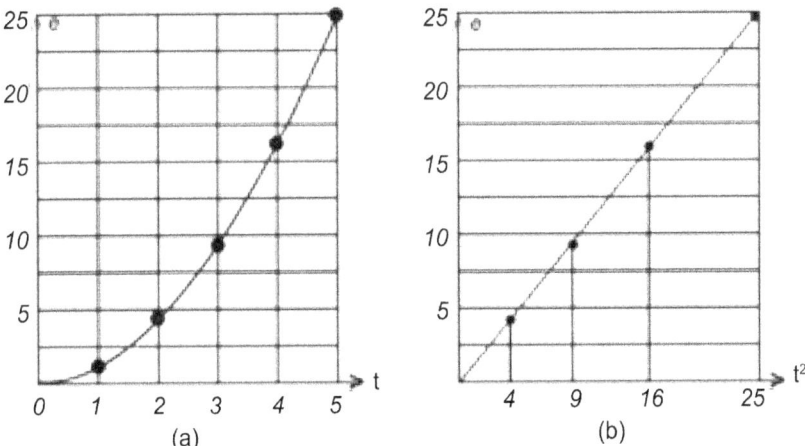

Figure 4: CARTESIAN REPRESENTATION OF THE FALL OF BODIES
(a) e vs t and (b) e vs t^2

From this data can be deduced variable speeds of bodies in free fall, their variations with time or their accelerations, and in general the elements of the new science called kinematics. It would be necessary to introduce a new kind of mathematics, called differential calculus, invented by Newton and Leibniz, to define in a general form the speeds and instantaneous accelerations, thereby establishing mechanics in no uncertain way.

The astronomical observations that suggested the trajectories of some celestial bodies with relationship to a fixed background that seemingly rotated on a daily basis around us, and the recent geographical discoveries that verified the hypothesis that our Earth was a great sphere, as were possibly other celestial bodies, brought us to a new understanding that established conclusively the current model of our solar system that we consider as "evidence" in our daily life. On the other hand, these observations should be compared with the earthly experiences of bodies in free fall that Galileo had established when describing mathematically the variation of their speed, which, contrary to the widespread belief originating with Aristotle, turned out to be independent of the weight of the body that was measured by means of the scale.

Galileo was persecuted for his celestial beliefs that contradicted the sacred books. He subsequently was forgiven, probably because his earthly studies favored the powerful princes who saw with good eyes the military applications of the new ballistic understanding. Anyway, the power of the church banished toward the north the new culture that flourished in the great synthesis of Newtonian mechanics, and established the new science.

5-2 FOUNDATIONS OF CLASSICAL MECHANICS

Isaac Newton carried out this great synthesis of astronomical and local observations by means of the mathematical symbolism of infinitesimal calculus, which made possible the formulation of a coherent theory that interpreted in a unified form the known experimental data, guiding the new mechanical science toward an unexpected development.

Newton's principles can be summarized as follows:

1. The law of inertia
2. Action and reaction law
3. $f = ma$

Let's now take a closer look at these principles:

1. The law of inertia says that a body in a state of rest or movement with constant speed on a straight line will remain in that state unless a force is applied to it. It establishes the concept of force as the cause for change in the dynamic state of a body.

2. The action and reaction law states that if a body A exerts a force on another body B, then B will exert an opposite force on A of

the same numeric value. It defines the force as a vector—that is to say, a mathematical entity that has a numeric magnitude, called a value, and a geometric, spatial character that is represented by means of a direction of an arrow. It is convenient to notice that the action and reaction forces are applied to two different bodies A and B and therefore they are not nullified, a case that may occur when several forces are applied to the same body and they combine according to the vectorial laws of the parallelogram. In this way, these first two laws also establish the mechanics for the state of equilibrium or statics.

3. The third law that is the basis for dynamics introduces the concept of acceleration, a, as the variation of speed with time, and relates it with the magnitudes of mass and force. Let us note that this formula is a vectorial relationship of the variables, f and a, which have a mathematical difference only in magnitude, even though conceptually they are very different. The importance of this difference led to new formulations for Newton's theory, such as those of Lagrange and Hamilton, which suppress the force but are, however, fundamentally equivalent, even though of greater deductive power.

Finally, Einstein, with his theory of general relativity, conclusively dethroned force from its pedestal, at least in dynamics. Nevertheless, neither the physicist nor the common man on the street has been able to remove this entelechy of the past from our daily language and therefore from our imagination. Worse luck have had other similar species, like the spirit or the soul, even when they remain in a cataleptic state. Generally, one forgets that in all of this lies the concept of cause, to which we have already referred and will have occasion to supplement ahead.

The definition of mass as a quantity of matter needs to be implemented, specifying how it is measured. If it is performed by means of the formula $f = ma$, we first will have to define force, and conversely. To avoid this cycle, we appeal to the traditional method of the scale to measure mass, perhaps the most important instrument of all time. In addition, we shall see it plays a fundamental historical role in chemistry, the origin of modern atomic theory.

The scale uses a static method; that is to say, it doesn't imply analysis of the movement of bodies, as opposed to Newton's formula that by definition is dynamic. This instrument compares weights—forces that we call gravitational that refer to the tendency of all bodies to fall toward the terrestrial surface. The concept of force is implicit in the measurement of mass on the scale, but this instrument doesn't measure movements that involve the variable time. Forces in statics are virtual; that is to say, their effects are not manifested. On the other hand, in dynamics, those goblins are known for their effects and not in themselves; we would say in colloquial language that "we don't see them, but that they are there, they are there."

So we may wonder: What was the novelty in Newton's theory, then? It interprets, in a more general form, Galileo's discovery that all bodies fall independently of their mass with the same speed; its originality doesn't properly reside in its formulas, which can be considered like a definition of force. Its originality is in a new mathematical technique, called infinitesimal calculus, which allowed the generalization of kinematics, transferring it to the celestial space when postulating the law of universal gravitation. This law expresses the force of attraction among the bodies, $f = G\, mM/R^2$, where m and M represent the two masses, R the distance between these, and G the gravitational constant.

If we conjugate the formula of Newton ($f = ma$) and that of gravitation, we can eliminate mass when assuming that the inertial mass is the same mass that appears in gravitation. This way we obtain:

$$a = g = G \cdot M / R^2$$

This justifies the use of the scale to measure masses, when expressing that the acceleration of gravity, g, is a constant in a given place on Earth (where the scale is), and the radius of the Earth, R, is given. In this way we establish that the same weights $(m_1) \cdot g = (m_2) \cdot g$, in the two arms of the scale, correspond to the same masses, $m_1 = m_2$. We find this result obvious at first sight; nevertheless, a comparison of gravitational forces with the electric ones that obey the law of Coulomb ($f = E\, q_1 \cdot q_2 / R^2$), in which the forces have a similar expression but depend on the electric charges (q) instead of the masses, illustrates the singularity of gravity, where the inertial masses are identical to gravitational masses.

These concepts of force and mass remained in their original form for two centuries during the time classical mechanics was developed. Hamilton presented Newton's formula in a novel, although equivalent, form, eliminating force and substituting it for energy that is considered in two ways: the kinetic energy and the potential. Kinetic corresponds to the capacity that a body in motion possesses of transmitting its energy to another system as they interact. It is expressed algebraically this way: $K = \frac{1}{2}\, (p^2/m)$, where $p = m \cdot v$ represents the momentum or quantity of movement and v the speed of the body.

The potential energy is symbolized by V and constitutes a property of a given space—for example, the gravitational field—where these forces are exerted, or the electric field if it is considered this way. We say that this value of V is a function of the space coordinates and time, which indicates the potentiality of the field producing energy on matter. We formulate it as V(x,t), where x represents the space coordinates invented by Descartes and t the time. In this way, we have Hamilton's formula, where H refers to the total energy of a system:

$$H\ (p,x,t) = K + V = \frac{1}{2}\ (\ p^2/\ 2m\ ) + V\ (x,t)$$

Hamilton's formulation has a big conceptual advantage when considering energy as a fundamental concept in physics; it also facilitates the manipulation and mathematical presentation when choosing as relevant variables the space coordinates and the momentum. We will have occasion to see their importance in quantum mechanics, which was formulated one century after Hamilton.

5-3 INFINITESIMAL CALCULUS

In astronomy, the expression of universal gravitation allowed Newton to calculate the trajectories of the planets in their movement around the sun, as a consequence of the acceleration, g, that is an inverse function to the distance, R squared, from the sun to each planet, independently as a first approximation of the mass of each one.

The importance and generality of Newton's work is directly related to the formulation of physical laws by means of a novel mathematical method that expresses the variations of physical magnitudes by means of "infinitesimal" variations—that is to say, by means of very small variations, as small as we wish them to be, which allow us to calculate their relationships extrapolated to finite limits. We define in this way the speed of a body by means of the relationship (dx/dt) between the variation of the traveled space, dx, and the time elapsed during that variation, dt, as the limit of this quotient when the space and time variations tend toward zero. This relationship that is denominated in a general form, derivate of x with respect to t, can be extended to any physical variable of any order; by this time and independently, this concept of derivate was formulated by Gottfried Leibniz, as a part of analytical geometry, which relates geometry and algebra by means of the Cartesian coordinates.

In Newtonian mechanics, the equation, f = ma, the acceleration, a = dv/dt, is introduced as the derivative of the speed, v, with regard to time, or as d(dx/dt)/dt, called the second derivate of space with respect to time. On the other hand, the force is expressed by means of an algebraic relationship that defines it with a relationship to the same space variables, f(x,t). This way we obtain an equation called a differential equation, because it involves, in addition to its magnitudes, their differential variations:

$$f(x,t) = m. (d(dx/dt)/dt)$$

The solutions of these differential equations are the fundamental objective of infinitesimal calculus, and imply at first the calculation of derivatives, which develop methods to find their algebraic expressions, that is to say, given the functions x = F(t), to calculate their derivate, dx/dt, and inversely by means of integral calculus to carry out the inverse operation, given the derivative of a function to find the corresponding function.

The importance of Newtonian mechanics doesn't reside properly in the equation that defines the force, f = ma, but in the introduction of a general method of calculation that makes it possible to obtain the trajectories—that is, to define the kinematics of bodies, by means of the differential equations adapted to each system or field of forces, as in the case of the gravitational or the electric field.

The expression of the laws of physics by means of differential equations constitutes the fundamental method par excellence to define relationships of the magnitudes that are considered relevant to describe natural phenomena. This is true to this day, in as much for relativistic physics as well as quantum physics, which have arisen and evolved in theoretical frameworks of greater generality, conserving Newton's and Leibniz's methods of infinitesimal calculus.

Newtonian mechanics constitutes the fundamental basis of classical physics, which was modified by the relativity theory that established, in the first instance, the so-called space-time, a mathematical artifice by which mechanics is integrated into a space of four dimensions, allowing the interpretation of mechanics when the relative speed of bodies approaches the speed of light.

5-4 RELATIVITY

Einstein's relativistic mechanics has two historical components: special relativity, which refers to mechanical systems called inertial (with constant speed, one with respect to the other), which are not subjected to acceleration, and those that are, which constitute the domain of general relativity.

5-4-1 Special Theory

Following the same logic as defined in Chapter 3, Section 3-2, in the presentation of the theory of relativity, it is convenient to distinguish the aspects of its conceptual origin and the mathematical methods that it uses. In special theory, the basic intuitive element is the relative character that time acquires to the observer, in a similar way as it happens with spatial dimensions. Newton's and Galileo's universal and absolute characteristic of time disappears, and acquires a local validity, integrating with space on equal footing.

This singular property of physical time is a consequence of the experimental fact that the speed of light is independent of the inertial system in which it is measured. These experiments were carried out for the first time at the end of the nineteenth century by Albert Michelson and Edward Morley at the Case Institute of Technology in Cleveland, Ohio. The speed of light was measured in perpendicular directions, a parallel one to the movement of the Earth around the sun and another perpendicular to this one, in the same place and at the same time. In another similar experiment, the speed of light was measured six months later when the Earth's movement was in the opposite direction with respect to the sun, always obtaining the same value for the speed of light.

The mathematical aspect of the special theory is deduced starting from two postulates:

1. The speed of light, c, is a universal constant.

2. The laws of physics are independent of the observer and are expressed in the same way in all inertial systems, that is to say, with constant relative speeds.

In its mathematical development, the idea of time as a relative dimension to the observer is introduced by means of simple transformations of matrices that, operating on the four coordinates of space-time, allow establishing their relationships when passing from one system of an observer to another. The algebraic terms of these matrices are deduced, making the corresponding transformations compatible with the postulates of relativity, and are equivalent to the algebraic relationships that Lorentz had deduced previously, and that Einstein stated not to know when his theory was published. Geniuses also have their weaknesses; they are made of the same dirt all mortals are made of.

Einstein's originality resides in his intuition regarding the nature of time that allowed for a new conception of mechanics, in contrast with Lorentz, who proposed unfruitfully the contraction or expansion of objects as well as time with their relative movement, looking for the cause of the invariance of the speed of light that was not compatible with the sum of the speeds or Galilean transformation.

5-4-2 General Theory

In general relativity, the fundamental idea arises directly from Newtonian mechanics, from an analysis of the singular character of gravitation that suggests a modification to the space-time of the special theory, which loses its Euclidian characteristic, giving place to a curved space that was inspired by the geometry invented one century before by Riemann. The shape of this space, that is to say, its geometry, depends on the presence of matter; in mathematical terms the coordinates of time and space depend as a whole on the mass present. The experimental fact that this gravitational mass is identical to the inertial mass in Newton's formula suggests that gravitation can be interpreted based on non-inertial coordinate systems.

The mathematics employed in the general theory is more elaborate than the matrix transformations of the special theory, and is called tensorial algebra. It basically preserves the same logic structure, using systems with relative acceleration that substitute the force.

Some idealistic experiments are useful to visualize the origin of the relativistic concept of gravity. If we imagine ourselves in an elevator standing on a scale, we will be surprised to notice that the scale will show a higher weight as the elevator starts up, until we reach a constant speed and the weight gauged returns to its original value.

If the elevator is going down, initially our weight diminishes, returning to the initial value at rest when its speed is constant. If we assume that the cable of the elevator breaks and we are in free fall, our stomach will indicate we are in a vacuum and the scale will gauge a zero weight; gravity has disappeared. These experiences agree with Newton's postulates; however, it is necessary to question how we could differentiate gravitational acceleration from that produced in a different way by means of, say, a reaction motor.

Let us move on to a less familiar thought experiment than the one Einstein used with his elevator, to a rocket in the current space era that is already common knowledge for children and young adults nowadays, even though less so for us older adults who still cling to habits that have been overcome. Let's imagine that we are on a space trip far away from any well-known star, without being able to refer our position to any other system. How will we be able to detect a possible acceleration that we may be subjected to at any given moment? If we assume that the origin of that acceleration is produced by the motors of the rocket, it will become evident in a similar form as in the elevator, by detecting the "weight," the inertia of any object, including the astronaut's own as he crashes into the walls of the ship.

On the other hand, we can assume that our ship is subjected to gravitational acceleration when we come closer to a black hole or to a star that we cannot see. In this case that gravity would be undetectable to the astronaut, even if he had Einstein's mind and NASA's laboratory. No weight would be noticed inside the ship; only when the unfortunate astronaut and his ship collided with the star or arrived at the black hole would the graveness of the situation reveal itself catastrophically.

In this example, we only notice the acceleration that gravity subjects us to in free fall if a crash occurs in which, expressed in physical terms, we pass from one system of measurement to another. If the two systems move at the same speed, it is said that these two systems are inertial. If the systems, the astronaut, and the black hole have a relative acceleration, we say the opposite—that the systems are noninertial. In both cases, we evidence acceleration when we reduce the system of the astronaut to the system of the ship or to the system of the black hole or star. Then what is the difference between the mysterious gravity and the thrust force of the motors?

Again we appeal to Einstein, who suggests we perform another thought experiment. Will there be some form to detect gravity before the catastrophe of the crash? By means of a sophisticated test in NASA's astral laboratory, we can make a geometric observation, measuring the trajectories of two objects that are located at a far enough distance so that those two lines converge toward a point located outside the ship that is no more or no less than the center of the black hole. In conclusion, gravity is detected by a geometric characteristic that causes two straight, parallel lines with relationship to inertial systems inside the ship to cease being so with respect to non-inertial systems, the ship and the black hole or the near star. In this way, the theory of general relativity detects gravity, by means of a space that has some different geometric rules from those that we are accustomed to in our local environment—that is, a non-Euclidian geometry where parallel lines intersect. Let us move on then to the mathematical language.

5-4-3 Some Mathematical Aspects of Special Relativity

General relativity and special relativity both employ mathematical techniques to relate the coordinates of different systems that correspond to observers located under different kinetic conditions:

a. In special theory, the two systems are inertial; that is to say, they move one with respect to the other with speed, v, which is a constant. For example, consider a peasant farmer who sees a train pass him as he stands very close to the moving train. The coordinates of an event with respect to the peasant farmer, whom we assume is at rest, are designated by a couple of numbers (x,t), and with respect to a passenger on the train (x', t'). The distances x, x' are measured on the direction of the train track, which is assumed straight, and the times t, t' correspond to the two observers who, when they passed in front of each other, synchronized their clocks on time zero. In geometric terms these initial conditions are expressed by saying that the origins of the two coordinate systems coincide; that is to say, x = 0, t = 0 corresponds to x' = 0, t' = 0 and the train moves with the passenger at a constant speed v on a straight line that coincides with the axis x. If three coordinates were used, then the values of the other two, y and z, would be zero as well as the speeds in those directions.

b. In the general theory, the two systems to consider are non-inertial; that is to say, they move with a relative acceleration, a.

The mathematical transformations that are used to relate these coordinates will be the basic tools of these theories and will allow us to obtain, starting from the general coordinates (t,x,y,z), the coordinates (t', x', y', z'). We will call these transformations operators, P, and express them this way:

$$P\ (\ t,x,y,z\ ) \rightarrow (t', x', y', z')$$

The operators for this theory are called tensors, and in their simplest form are matrices of equal dimension to those of the coordinates. These transformations can also be expressed in algebraic form, and are denominated in the special theory as Lorentz's transformations.

The results of these formulations or calculations can be represented geometrically, using the space coordinates in a line or in a horizontal plane, and time in the vertical one (see Figure 2). The general case of three space coordinates is not useful for a graphic representation, because our senses don't allow us to imagine four dimensions when including time. This class of diagrams is denominated in the space-time of Minkowsky, who was Einstein's professor and invented them in 1908, after the publication of special relativity in 1905. Undoubtedly, this representation has been of great utility to facilitate the interpretation of the special theory and for the formulation of the general theory that was published by Einstein in 1918; in a similar way, Descartes's coordinates were fundamental for the understanding and development of Galilean and Newtonian mechanics.

Minkowsky's space is a very particular one; it doesn't refer to trajectories of particles in Newtonian space coordinates (x,y,z), where time is a parameter that occurs independently of the observer. This new space represents events that happen at a certain time in a position in space, with a special characteristic that the time coordinate is expressed in space units, when multiplying by c, the speed of light $(z = c.t)$. A line in the diagram of Minkowsky represents the history of a given body.

An algebraic development employing two-dimensional matrices (x,t) allows us to find, in a simple and rigorous form, the transformation operators of coordinates for inertial systems that uphold the postulates of special theory:

a. The speed of light is constant.

b. The formulation of physical laws is independent of the coordinate system.

These operators that are equivalent to the Lorentz transformations allow for very compact formulation of the equations of relativistic mechanics, such as, for example, the formulation of the famous formula:

$$E = mc^2$$

This formula is calculated when expressing the "covariance" of the law for the conservation of linear moment in inertial systems; that is to say, by means of the operators that transform the vectorial speed among the two coordinate systems, when employing the principle (b) expressed in special theory.

In a similar way the "invariance" of the vectorial position in a four-dimensional Minkowsky space that we have referred to in chapter 4 is demonstrated:

$$x^2 + y^2 + z^2 - (c.t)^2 = d^2,$$

where d is the longitude of the tetra-vectorial position that is a constant in inertial systems. We can consider this invariance the new theorem of Pythagoras for space-time.

The Lorentz transformations were used before the formulation of the theory of relativity to explain the constant nature of the speed of light independent of the inertial system in which it was measured, by means of the supposed contraction of the length and the intervals of time with the relative movement. The importance of relativity is expressed in the introduction of time as a fourth dimension of space on an equal footing to the Cartesian coordinates of space. Minkowsky's space adds to time the character of an imaginary coordinate (i.c.t), where $i = (-1)^{1/2}$, which makes its square, $(i.c.t)^2 = -(c.t)^2$, giving to the invariance of the position vector an identical mathematical expression to the relationship of Pythagoras.

5-4-4 Social Impact

The current impact of the theory of relativity on the average person may seem surprising, being that it's a theory based on concepts so abstract and occult. The special theory was published in 1905 and was received with skepticism and distrust in the scientific community. Let us remember that its author was rewarded with the Nobel Prize for his studies of the photoelectric phenomenon that not only had importance for the incipient quantum theory of the time, but was also directly related to promising applications. Certainly, beginning with the Prize's founder, the engineer Alfred Nobel, the politics of selection of Prizes in science has been mainly based on the candidates' experimental aspect and application character, and not necessarily on the theoretical point of view.

For relativity to be accepted by the scientific community and later have such wide coverage in all the intellectual and journalistic environments, it was necessary to have a spectacular confirmation of its experimental predictions. Of these, the most published in connection with the theory of gravitation was the experiment carried out in 1919, on the deviation of the luminous trajectory of a star that coincidentally could be observed as it moved close to the sun during a total eclipse. The relativistic calculations based on the effect of solar gravitation on the ray of light coincided with great accuracy with the angular deviation of the stellar position observed during the eclipse. On the other hand, eclipses historically have had great mythical meaning, and it was natural that this experiment had great journalistic diffusion.

Another aspect of relativity that has contributed notoriously to its popularity refers to the equivalence of the energy and mass that is a consequence of the special theory. Forty years and the unfortunate explosion of the first atomic bomb in Hiroshima were needed for the equation $E = mc^2$ to go around the world, and even begin to appear in street graffiti. Less well-known is the importance of this equation in explaining the origin of solar and universal energy that makes possible our existence.

Einstein is the best-known scientist of all time. We all have great admiration for his scientific work; however, his theory of time remains far from our imagination, and only by means of the mathematics and geometric representation of Minkowsky can it be expressed appropriately. Psychological time and space are valid individually or privately, in contrast with the physical symbolism that is universal, or public; this is because we don't possess the gift of ubiquity, of being located simultaneously in two places. Only the persistency of some experimental facts has made it possible for our local world of senses to transcend, by means of relativity, to the origins of the universe.

5-4-5 The Shape of Space-Time and the Origin of the Universe

Recently Hawking,[1] Penrose, and other authors have popularized important aspects of the theory of relativity in relation to the origin of the universe and its cosmic evolution. Occult terms like "curved space," "fourth dimension," "big bang," "black holes," and "trips in time" have passed into the domain of popular language, influenced by science fiction novels and videos. It is not an easy task to translate the abstract mathematical symbolism of the formulations within physics to common language. The use of the graph, as a bridge between these two symbolic universes, turns out to be a powerful means for acquiring understanding that is not limited only to books of scientific divulgation, but which also has been undoubtedly used by the creators of science to "visualize" their extraordinary theories.

Cosmology has been a historical constant in the development of human thought. The relativistic theory represents the frontier of thought, in the understanding of the universe.

The integration of space and time in a coherent group allows us to visualize the universe as a finite and limitless space of four dimensions, to the way of a spherical surface, in which there is no privileged place.

---

[1] Stephen W. Hawking, *The Universe in a Nutshell* (New York, Bantam Books, 2001).

This spherical surface is closed and finite; it doesn't have limits. If we were two-dimensional beings, we could not leave that jail that would give us the apparent sensation of freedom, in a plane that extends in all directions. There is nothing external in this curved universe, in contrast with the Euclidian infinite universe that follows the game of the Russian dolls that hold each other up indefinitely.

In any direction one may choose to go, the universe is homogeneous, not in a localized way but at a macrocosmic scale: the electromagnetic radiation we receive has its origin in all places at all times. The universal "background" of microwaves has the same intensity and distribution of frequencies in all directions of space wherever in the universe we may be located. It is the "noise" that arrives until "now," in any "given now" that is prolonged and has been prolonged indefinitely, spreading since the big bang or "singularity" in which the universal expansion began, up to "here" in the planet Earth that is equal to any "here," after interacting with the whole universe, during all its finite history.

Electromagnetic waves that we receive in their entire spectrum, from the cosmic rays to microwaves, provide us simultaneously with information about time and space. To complement the image of the spherical surface, useful to conceive the cosmic space, we use the Minkowsky representation (see Figure 5). We imagine our virtual trip through time,

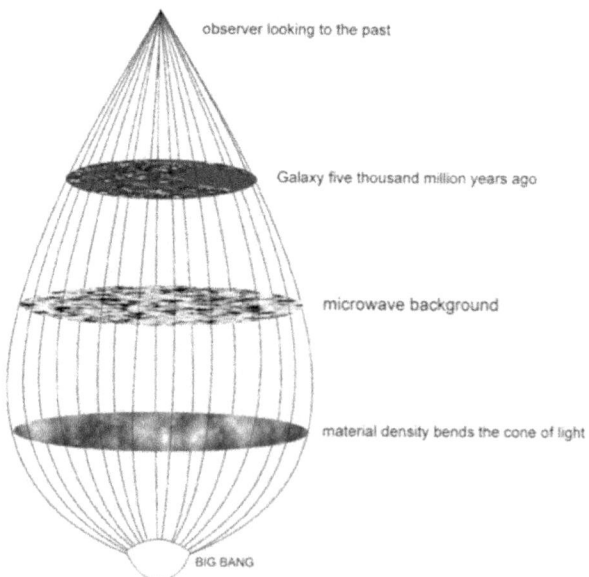

Figure 5: **CONE OF LIGHT OF OUR PAST**

looking to the past from our present time, located in the origin of the cone of light that we receive from the singularity of the "big bang," which bends in its primitive stage inward, due to the high density of our birthing universe. This representation also shows us stages closer in time, corresponding respectively to the background of microwaves and galaxies in their evolution. The physical history of the universe has a fascinating effect on the imagination, and exerts an important ethical and social influence.

5-5 EPISTEMOLOGY OF CHEMISTRY

With this title we want to present the nature of knowledge related to material diversity, by means of the study of the historical processes that have created it. In this analysis we will keep in mind the evolution of philosophical and scientific thought, as well as technological advances and pertinent experimental methods.

5-5-1 Alchemy, Philosophy, and Science

The historical origin of the word *chemistry*, as well as its equivalents in other European languages, such as *química* and *chemie,* derives from the word *alchemy,* which in turn comes from the Arab article *al* and the Greek word *chymeia,* for which the fusion of metals was designated, probably for its relationship with the term *kypros,* which referred to the metal copper and to the island of Cyprus from where it came.[1] This terminology and its historical evolution illustrate the importance that obtaining metals had in the science we today call chemistry.

Metals were historically important in technological and military development, and today are used for designating great historical eras. In the past metals were considered symbols of power when relating them with stars: the sun with gold, the moon with silver, Saturn (in Greek, lead), with the slowest of the planets that also symbolized gods, such as Jupiter and Venus. Undoubtedly, metallurgy, the origin of alchemy, represents one of man's biggest technical achievements.

During the Middle Ages, a good part of the Greek culture arrived in Europe through Arab translations of works from very diverse branches of old Mediterranean and Eastern civilizations that understood philo-

---

[1] A. García and J. Bertomeu, *Name the Matter: A Historical Introduction to Chemical Terminology. Nombrar la materia. Una introducción histórica a la terminología química* (Ediciones del Serbal, 1999).

sophical and mathematical classical disciplines, as well as empiric clusters of knowledge, such as alchemy, where hidden procedures of practical importance were expressed by occult symbols and mythological denominations.

One of the obsessive objectives of the alchemists, to the extent of becoming almost their only activity, was the search for the transformation or transmutation of matter into gold, which anecdotally was designated as the search for the "philosophers' stone." It seems surprising that Isaac Newton, founder of the new physical science that revolutionized the knowledge of nature by means of mathematical laws for the movement and gravity of bodies, did not glimpse that, in the empiricism of alchemistic knowledge, there were technological procedural elements present capable of analyzing from a quantitative angle the origin of the diversity of matter. Newton's genius extended to multiple physical aspects, like the corpuscular theory of light, placing him centuries ahead of his time; however, in connection with the composition of matter, he clung to the alchemist methodology that studied the likeness among different substances, only looking for their transformation, without introducing quantitative elements in the experimentation and sharing the utilitarian interest for the search for transmutation—an attitude that doesn't favor the scientific position that values knowledge for its intrinsic value.

The study of "chemical affinity" was not the appropriate approach to begin the difficult process that could lead to a coherent theory for the diversity of nature. In the seventeenth century, the knowledge of alchemists had been summarized in tables of affinity like that of Étienne François Geoffroy (1672–1731), which is presented in Figure 6, taken from the book of A. García, already cited. These tables, organized into columns, refer to the relative reactivity of some substances with others. For example, the sixth column of the chart is headed by alkalis or ash, followed successively by vitriol (sulfuric acid), nitric acid, sea salt acid (hydrochloric acid), the spirit of vinegar (acetic acid), and sulfur. This indicates the order in which strong acids displace weak ones in their compounds with ash—that is to say, with their salts.

Such empiric data continued being published in a similar way, as the number of discovered substances was enlarged, without advancing in a formulation of theories that described their properties. Initially new theories aimed to look for models that indicated the composition of matter and its parts. Nevertheless to have a better understanding of these early theories, it is convenient to keep in mind that the Aristotelian Greek thought that prevailed in the time of the Renaissance referred to the so-called

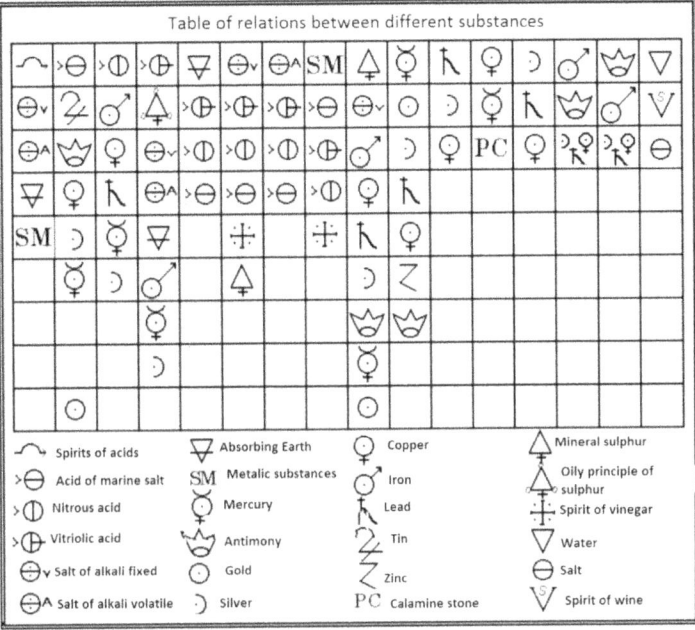

Figure 6: GEOFFROY TABLE OF AFFINITIES

essences, or fundamental qualities of matter, and not properly to its parts or constituents. As a consequence, an aspect that intrigued investigators was related to the so-called spirits of the substances, which in some cases referred to an "air" or to estates like "heat," which indicated the capacity to produce fire to which was given the Greek name of *phlogiston*. The confrontation of all these ideas gave rise to a new experimental approach, introducing the measure of weights of the reactive substances and making possible the formulation of the new chemical science.

Experimentally, this new direction for the investigation of material diversity was guided by studying the so-called *airs*, a word that evolved into *gas*, meaning chaos. Its properties were measured by its density, based on the improvement of the scale and on the invention of recipients and appropriate procedures to confine them. On the other hand, from the theoretical or philosophical point of view, evolution of the hypotheses of the composition of matter began with the ideas expressed by the Greeks Leucipo and Democrito (fourth century B.C.), by means of the word *atom*, which is formed by the prefix *a*, indicating negation, and *tomo*, meaning part. Thus, the word designated its belief that everything is composed of these atoms that are indestructible or indivisible and move in the empty space that exists between them. This understanding of matter, which we find close to current science, also speculated about the diversity of matter, when referring to the kind of atoms whose differences were enunciated by their form and size and further differentiated by their heat and weight.

However, Greek thought didn't differentiate clearly between facts or empiric properties, derived directly from observation, and logical and speculative arguments, in such a way that would influence philosophers such as Plato and Aristotle, who finally had a much bigger impact than their predecessors on the theocratic philosophy of medieval Christianity. Aristotle stated that all bodies were formed by four elements or principles; air, earth, water, and fire, which designated qualities like cold, hot, humid, and dry, whose interactions in different proportions explained the material diversity and their transformations. (See Figure 7, from the book of Michael Maier, *Del circulo physico quadrato*, taken from the book of A. García already cited.) This understanding and that of the atomists, which were to a certain extent antagonistic, were developed in parallel in the pre-scientific atmosphere of the Renaissance, giving place with their confrontation to the consolidation of chemistry.

Aristotle's authority, which had been undermined by the new physical science of Galileo and Newton as the theory and experimental data harmonized, still maintained its domain in the world at that time, perhaps

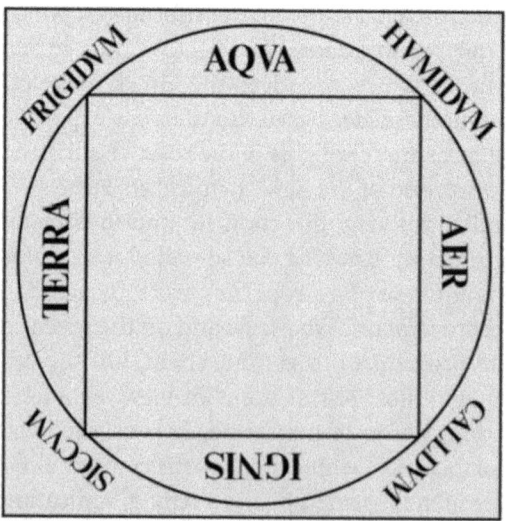

Figure 7: THE FOUR ARISTOTELIAN ELEMENTS: WATER, EARTH, AIR, AND FIRE. THE QUALIA: HUMID, DRY, COLD, AND HOT.

because his teachings were closer to our immediate impressions of reality that show to us simultaneously plurality and unity. In a certain way the controversy was about semantic interpretations related to metaphysics: What aspects define our sensations? Are these the qualities we assign to the substrate or the substance? Or is matter its cause? Or is our mind, the spirit, the only reality?

Centuries passed before this ancestral controversy was decanted; we suggest the reader analyze the evolving arguments in this respect in Chapter 6, "Matter and Spirit." In any event, the scientific position called "positivist" little by little was gaining ground in this territory, offering a different option that avoided this ideological confrontation, focusing instead on the so-called principle of objectivity to which we have already referred. According to this principle, theories or mental models should adapt and be modified so that they agree logically with experimental data, and not vice versa, when assuming that those are nonmodifiable, as mathematical truths are. The foundation of chemistry is a good historical example of that evolution.

5-5-2 Importance of the Historic and Semantic Analysis

Frequently, when studying the history of science, terms or words misplaced of their temporary meaning are used, forgetting that limitations of language are not solely geographical, but also historical. Words like *physics* and *chemistry* cannot be used in the contemporary sense, without considering previous or simultaneous times of basic development of these sciences. We should not forget that the divisions of knowledge are simply conventions that we use to symbolize their evolution. In this way, as the environment and depth of knowledge is enlarged, philosophy can pass as a part of science, and within this, the term *physics* advances in lands of what we call chemistry or biology, and—why not?—in psychology.

On the other hand, the classifications of knowledge frequently refer to the professions and not to science itself; these economic characteristics also evolve. These reflections are appropriate for measuring the meaning of the dominant opinion of contemporary physicists in connection with other natural sciences, as such expressed by Isabel Stengers,[1] as follows: "Chemistry, is it a branch of physics characterized by its approximate or elementary nature of its intellectual operations and practices at the same time? Physics is understandable, chemistry is learned."

---

[1] M. Serres, ed., *History of Sciences. Historia de las ciencias* (Ediciones Cátedra S.A., 1998).

Contemporarily, we can say that the word *physics* is used mainly to designate the area of natural sciences that uses deductive mathematical logic, and that such sciences as chemistry, biology, and psychology, as they study more complex facts, enrich themselves initially in a qualitative field, progressing at least in part to mathematical physics, which evolves in its increasingly more rigorous nature toward new methods of analysis. Biology today, influenced by physical chemistry, acquires experimentally and theoretically an extraordinary dynamism that reminds us of passed eras in chemistry, a science that has already acquired greater maturity.

Going back to the epistemology of chemistry, in its voyage from the empiricism of alchemy toward science, we cannot use terms like physical or chemical properties simply because those differences didn't exist before the basic theory of chemistry was established. That theory is founded on the assumption that matter is formed by fundamental units that characterize its global properties and explain its diversity and transformations. The properties and definitions of these units evolved as experimental knowledge advanced, and were perfected by their interpretation in a qualitative and quantitative form. Therefore, consider as an example, in a historical context, the transformation of a substance like water when passing from solid to liquid or vapor, in contrast with their formation starting from two "airs" or gases recently discovered. We cannot designate with authority one or the other as physical or chemical transformations, terms that didn't have a meaning at that time.

5-5-3 The Objectivity Principle Foundation for Science

Experimentation with the airs, later called gases, had its origin in the observation of the phenomena of combustion and in animal breathing. When George Stahl[1] (year 1700) proposed a theory to interpret this phenomenon, he assigned to the bodies able to produce it: an "igneous principle" called phlogiston that was considered in the same way as Aristotle's element of fire, as a fundamental quality of nature. The novelty of this theory of *phlogiston* and its historical importance reside in its transferring this hypothesis from the exclusive metaphysical world to the experimental one, allowing then the introduction of the scientific method to ideological controversies.

By extension, such phenomena as the breathing of animals and the transformation of metals into earthly substances were assimilated into

---

[1] Isaac Asimov, *A Short History of Chemistry* (New York: Doubleday, 1965).

combustion. In this way, when metal becomes rust, it loses its phlogiston; inversely, earthlike substances recovered their phlogiston, reverting again to metal by means of their interaction with coal, which possessed the so-called igneous principle. Natural air was not granted the possession of phlogiston, but rather it was considered a vehicle to transport it from one place to another. Perhaps the reason for this curious assessment was the loyalty to the Aristotelian authority that considered air an element of nature with similar status as fire.

In opposition to the Aristotelian theory of matter, called *hyle*, which considered its properties, or *qualia*, as entities or principles intrinsically responsible for the natural phenomena (see Figure 7), was the theory of atoms, which proposed that particles or indivisible parts possessed individually the properties of the whole. The differences between these two theories seem very subtle; however, the theory of the atomists proved more appropriate and fruitful, giving place to the belief or intuition that a limit existed for the analysis or subdivision of bodies, from which further subdivision could not be arrived at, and that this analysis should be performed in the laboratory, so these substances could be found and defined, though perhaps provisionally. Subsequently these simple substances were denominated "elements," a term coming from the Latin translation of the Greek word *stoicheion*, from which the word *stoichiometry* was derived, referring to the proportion of elements or simple substances that participate in the constitution of a compound substance.

The isolation starting from natural air of an element or "air" that had properties different from the whole, was an experience of extraordinary importance; much greater, who would have believed it, than the eventual alchemist dream of the search for the philosophers' stone. Paradoxically, the elements that chemistry postulated in the XIX century, became atoms that did not result indestructible, and the science already named physics, has been able to investigate in its constituent parts, proving experimentally the possibility of their transmutation. Greek atoms are now nuclear particles, and again, the hypothetical "quarks" have been postulated as fundamental, returning to the atomists.

A. THE AIRS AND THE PHLOGISTON

Let us return to the seventeenth century, when Joseph Priestley introduced the expressions "phlogisticated air" and "dephlogisticated air" to designate different behaviors of these parts of the natural air, with regard to the combustion of bodies. In the air without phlogiston, the substances lit their combustion, while in the air with phlogiston, combustion could not be sustained, nor could animal life. Some particulars of these experiences correspond to observations where airs with different properties were obtained, which are distinguished for their important contribution to the foundation of chemical theory (see Table 3).

NUMBER	ORIGIN	OLD NAME AUTHOR, YEAR	MODERN NAME
1	Heating limestone	"Fixed air or wild" Black, 1766	Carbonic gas
2	Coal combustion Air and separation	"Phlogisticated air" or azote Rutherford, 1772	Nitrogen
3	Mercury combustion Decomposition by heating "terra" or oxide	"Dephlogisticated air" Priestley, 1774	Oxygen
4	Metal dissolution with acid	"phlogiston"	Hydrogen
5	Hydrogen and oxygen combustion	Water	Water

Table 3: OBTAINMENT OF AIRS

For a better illustration in this respect, consider some aspects from these experiments as well as properties of the airs discovered:

1. The production of carbonic gas, or carbonic anhydride, by heating limestone, calcium carbonate, has importance in itself and also easily allowed the measurement of the loss of weight from the limestone when becoming "live lime," calcium oxide, producing an air that has the same properties of the one obtained (according to numeral 2 in Table 3) by combustion of coal. This "fixed air"' can be fixed back (thereby its name), inverting the reaction:

   Limestone + heat = fixed air + live lime or alkali

2. The first stage combustion of coal gives us phlogisticated air, or azote, in a second stage, when separating the fixed air obtained from natural air that was used in a closed recipient to burn coal, by means of the inverse reaction explained in numeral 1 in Table 3.

Schematically, obtaining of the azote or nitrogen proceeds as follows:

First stage   natural air + coal = fixed air + phlogisticated air

Second stage   fixed air + phlogisticated air + live lime = limestone (solid) + phlogisticated air

This phlogisticated air, also called azote (from Greek, meaning "without life"), suffocated animal life and impeded combustion.

3. Mercury was used by chemical pioneers such as Priestley to substitute water as a means to seal recipients in the so-called "pneumatic box" where the airs were collected. The advantage of mercury resides in its stability; it neither evaporates nor combines with dissolving airs, such as fixed air. Incidentally, however, it was found that mercury produced with natural air, as other metals did by means of combustion, a reddish earth that had the property, in the absence of air and by means of the action of concentrated solar rays from a magnifying glass, of reverting again into mercury, producing an air that had the opposite properties of the phlogisticated air or nitrogen (see Table 3, reaction 2), bringing life to combustion instead of snuffing it out and aiding breathing instead of suffocating it. The following outline summarizes this reaction that for the first time created oxygen, which its discoverer called "dephlogisticated air."

First stage   mercury + natural air = earth (mercurious oxide)

Second stage   earth or rust + light = mercury + dephlogisticated air (oxygen)

B. IMPORTANCE OF THE QUANTITATIVE FACTOR

Antoine-Laurent de Lavoisier in 1780 learned all these important experiments, repeated them carefully, and added others of his own genius, introducing in his analysis the quantitative factor, by means of measurements of the weight of the reactant components and resultants. From an experimental point of view, not only did he perfect the scale, increasing its precision, but also devised new separation and isolation methods for the gases, indispensable for measuring their weights. In this way he built a technology that had been initiated a century before in 1622 by Robert

Boyle, who in part at least can be considered as the initiator of scientific atomic theory. The law that takes his name establishes the relationships between pressure, p, and volume, v, of the gases:

$$p.v = constant$$

This law was found by measuring the pressure exercised by a column of mercury by its length on the gas confined in the same glass tube closed on the other end. This law gave place to the introduction of measurement in the analysis of phenomena of what today we call chemistry, and it can be considered as the origin of physical chemistry.

Take into account that the fact of measuring properties of the airs constituted a challenge to the prevalent thinking that considered air as a spirit, in opposition to water and earth, which represented solidness, or what is tangible. Boyle was a pioneer of the atomist theory that explained properties of the air, assuming that it was comprised of elementary particles, the atoms, located at variable distances from one another according to the pressure that was exerted on them. In accordance with these ideas, airs differed from liquids and solids, in the existence of space between the atoms that gave them the possibility to vary their volume. In this context airs should weigh the same as solids and liquids, being differentiated by their smaller density. In the same fashion, the transformations that today we call changes of physical state, such as the liquid water that becomes air, or what we today call "vapor," is explained by an increment of mobility of its atoms that distance themselves from one another, and not due to a change of its "essence" when becoming "spiritual."

Lavoisier's genius led him to create a synthesis of all these ideas and experiences, establishing by means of measures the law of the conservation of mass (weight) in all the observed transformations. Thus the shades of the phlogiston vanished: the dephlogisticated air opened the way to oxygen that bonded with bodies like metals in combustion to produce an earthlike substance that increased its weight in the measure that oxygen bonded to it. In a similar way, water that evaporated when heating it could become liquid again by cooling it without variation of weight.

On the other hand, the Aristotelian understanding was disappearing: Water was no longer an element; it originated from the combustion of an air (Table 3, numeral 4), that was believed to be the phlogiston (later called hydrogen, meaning it produces water with the new oxygen). The new theory defined, as basic entities of matter, material units called molecules (from Latin, meaning "small particle") that constituted the so-called "simple"' or "elemental" substances, defined as such provisionally while an appropriate way to decompose them was not found.

C. PRINCIPLES OF CHEMISTRY: SIMPLE AND COMPOUND SUBSTANCES

For a broader illustration, it is convenient to transcribe some of Lavoisier's expressions from his fundamental book *Elements of Chemistry*,[1] published in 1789:

> All we can say about the number and the nature of the elements is limited, in my view, to purely metaphysical discussions; they are indeterminate problems for which infinite solutions exist, of those which probably not one is in total agreement with nature. Therefore I will be satisfied with saying that if with the name of elements we seek to designate as simple and indivisible molecules that form the bodies, it is probable that we don't know them; that if on the contrary, we tie to the name of elements or of principles to the bodies, the idea of the last term that the analysis arrives at, all substances that we have not been able to break down by any means or method constitute elements for us; it is not that we cannot affirm that these bodies we consider simple are formed in turn by two or more principles, but, since these principles are never separated, or rather, since we don't have any method to separate them, they act in what concerns us, as simple bodies and there is no reason to assume that they are compounded, except if experience demonstrates this way.

By way of summary, we can say that the work of Lavoisier established the principles of chemistry this way:

1. It establishes the conservation of mass in the transformations of matter.
2. It defines substances, simple and compound.
3. It establishes a nomenclature to designate compound bodies based on simple ones, indicating their composition.

The new chemical nomenclature proposed provides an appropriate symbolism to express the importance of the composition of compound

---

[1] Serres, *Historia de las ciencias.*

substances, in connection with simple ones. Lavoisier proposed in this way a list of thirty-three simple substances or elements that were classified according to their reactivity:

> Substances belonging to the entire natural kingdom that can be considered as elements of bodies:
> > Light and caloric
> > Oxygen that produces acids, vital air
> > Hydrogen that generates water
> > Azote, inert gas, without life
>
> Simple, nonmetallic substances, capable of oxidizing and acidifying:
> > Sulphur or sulfur
> > Phosphorus
> > Carbon
>
> The known metals
>
> The earthlike substances such as magnesia, alumina, barite, silica, or sand
>
> Radicals such as "muriatic," "fluoric," and "boric"

Of these substances, classified as simple, only the earthlike substances we know are not. We don't know clearly to what the radicals refer, even when their names probably indicate hydrochloric, fluorhydric, and boric acids. As for the inclusion of light and caloric in the chart, this is reminiscent of Aristotle's principles.

These charts allowed for the expression of compound substances, such as the metallic oxides, formed from metal and oxygen; the fixed air from coal and oxygen; and the salts, such as the sulfates and metallic carbonates, formed from radical acids originating from sulfur, or from carbon and oxygen (meaning it produces acids), by means of names that expressed their composition.

Lavoisier's work was the foundation of chemistry, and its improvement and consolidation began by means of the experimental advance of analytical methods seeking to enlarge or diminish the list of simple substances and to obtain new compound substances, expressing their composition or stoichiometry by means of numeric symbolism or formulas. In this process, we see the sketching of a new direction that gives continuity to the ideas of Boyle, incorporating in the science, already called chemistry, elements similar to the Aristotelian principles, light and caloric, accepted by Lavoisier and paving a relationship with concepts of Newtonian mechanics such as energy.

5-5-4 Chemistry: The New Science

A. THE ATOMIC HYPOTHESIS

We can compare the development of chemistry, starting from 1800, when the atomic hypothesis was adopted, to the solution of a crossword puzzle. However, the chemical pioneers faced an additional difficulty: Not only did they have to unite well-known pieces in the appropriate form to reveal a figure or a word, but they also had to discover simultaneously which fundamental pieces built a harmonic group with the diversity of nature. The atomic hypothesis was the key that provided the theoretical framework to order coherently the experimental data that initially was reduced to the analytic results of the composition of matter, symbolized by means of so-called empirical formulas.

This method turned out to be appropriate mainly in the natural kingdom of the so-called mineral substances, where the laws of stoichiometry were enunciated as the law of definite proportions, which defines the constancy of the proportion in which simple elements participate in the constitution of a compound substance. This law complemented the law of multiple proportions, which refers to cases where two or more simple substances can combine to form compound substances, in proportions that differ in relationship of integers, indicating the fact that simple substances behave as indivisible units that remain as such when uniting in binary form or of a higher order, when forming a compound. Implicitly these "laws" express the difference between a chemical compound and a mixture of various substances whose composition is arbitrary, indicating, in turn, the strength of the chemical bond of its constituent elements. This suggests the importance of the concept of "affinity" held by the alchemists, which remained unexplained, waiting for new physical evidence related to the "caloric" or the energy of Newtonian mechanics.

B. CHEMICAL FORMULATION AND THE ATOMIC WEIGHTS

In the evolution of knowledge, symbols have always played a fundamental part. Just as in mathematics, the decimal system of numbers replaced the system of representation of the Romans, in a similar way, the pictorial symbols of the alchemists were simply replaced by letters of the Latin alphabet, designating the elementary substances, and by formulas, designating compound substances. John Dalton (1766–1844) and Jöns Jacob Berzelius (1779–1848), around the same time, symbolized atoms by means of letters wrapped in circles: H, O, C, S, and A referred to the atoms of hydrogen, oxygen, carbon, sulfur (sulphur, *soufre*), and azote (nitrogen), respectively, while the compounds were represented by the association of the atomic symbols, or formulas, indicating numerically by means of indexes the proportion in which they composed the molecule, based on atomic weights for each one of the atoms. Some of those formulas proposed by Berzelius and Dalton are presented in Table 4.

FORMULAS PROPOSED BY BERZELIUS	
NAME	FORMULA
Ferric oxide	Fe_2O_3
Potassium nitrate	KO, A_2O_5
Cupric sulphate	$CuO, SO_3 + 5HO$
Water	H_2O

FORMULAS PROPOSED BY DALTON	
Carbon monoxide	CO
Carbon dioxide	$OCO (CO_2)$
Water	HO

Table 4: FORMULAS PROPOSED BY BERZELIUS AND DALTON

The numeric indexes in the formulas are calculated based on the atomic weight of hydrogen, which was designated from the beginning as a unit representing the "lightest" gas, except for indexes in the formula for water, which Dalton assumed to be a binary molecule, in contrast with Berzelius, who supposed it ternary. To present with more clarity arguments for or against these formulations, we will use for the case of water the following stoichiometric equation:

$$m(H)/m(O) = Pat(H)/Pat(O) . n(H)/n(O)\hat{} \qquad (1)$$

where m represents the reactionary masses of the two gases to form water, n designates the indexes employed in the formula for each atom, and Pat the atomic weights. Assigning to the atomic weight of H the value 1, assuming the relationship of indexes and measuring the relationship of masses, we can calculate by means of the previous equation the atomic weight of oxygen. Other atomic weights were calculated using compounds of those elements formed with hydrogen and oxygen. Table 5 shows some of the atomic weights proposed by Stanislao Cannizzaro (1826–1910), Berzelius, and Dalton, compared with the actual ones taken from A. García and J. Bertomeu.

ELEMENT	ATOMIC WEIGHT, DALTON (1808)	ATOMIC WEIGHT, BERZELIUS (1828)	ATOMIC WEIGHT, CANNIZZARO (1860)	ATOMIC WEIGHT ACTUAL
Hydrogen	1	1	1	1.008
Oxygen	7	16	16	16.00
Nitrogen	5	14.2	14	14.01
Carbon	5.4	12.2	12	12.01
Potassium	35	78	39	39.01
Mercury	167	202	200	200.6

Table 5: ATOMIC WEIGHTS

C. THE HYPOTHESIS OF AVOGADRO AND THE MOLECULAR WEIGHTS

Referencing Table 5, we see that if we multiply by 2 the atomic weights proposed by Dalton, we obtain approximately the atomic weights of Berzelius; this is because Berzelius proposed the formula for water as H_2O, while Dalton proposed HO, and this served as the basis by which the other atomic weights were calculated. The reason for this difference is possibly due to the influence of Amedeo Avogadro and Joseph Louis Gay-Lussac, who used the formula of Berzelius for water as the base of theoretical and experimental considerations. At first it would seem that the reason for this difference, as we know today, is that oxygen and hydrogen molecules are diatomic at ambient temperature, and therefore the reactionary masses that are referred to by the experimentalists of that era were molecules and not atoms; however, the relationship of the molecular weights between hydrogen and oxygen is the same one as the atomic one.

This discrepancy between the formulas for water was a key point for the evolution of atomic theory; Avogadro[1] proposed his famous hypothesis, which we transcribe literally:

---

[1] Jefferson Hane Weaver, *The World of Physics* (New York: Simon & Schuster, 1987), G5, 621.

The primary hypothesis that arises in this context, and seemingly the only one acceptable, it is the assumption that an integer (N) of integral molecules, in (all) gases, is always the same for equal volumes or always proportional to the volumes (V).

Algebraically, we can express this hypothesis in the following way, with K being a constant:

$$N = K.V \qquad (2)$$

And the density d(i) of a gas (i) according to atomic theory and to formula (2):

$$d(i) = (N/V).(Pat)i = K.(Pat)i \qquad (3)$$

Or, in words, the density of a gas (i) is proportional to its atomic weight (Pat)i or, more appropriately, to its molecular weight.

Avogadro, according to his own memoirs, proposed his famous hypothesis, keeping in mind, among other factors, Gay-Lussac's experimental findings "that gases react in simple proportions in volume, as well as producing gases which are also related simply to its components." Definitely, in the case of the formation of water starting from hydrogen and oxygen, it was found that two volumes of hydrogen react with one of oxygen, to produce two volumes of water, measured at the same pressure and temperature, as follows:

$$2 \text{ (vol. Hydrogen)} + 1 \text{ (vol. Oxygen)} = 2 \text{ (vol. Water)}$$

The fact that in this reaction, as well as in others, like the formation of ammonia,

$$3 \text{ (vol. Hydrogen)} + 1 \text{ (vol. Nitrogen)} = 2 \text{ (vol. Ammonia)},$$

a decrease of volume also takes place, indicates, according to Avogadro's hypothesis, that in these reactions a decrease in the number of molecules occurs. Consequently it is inferred that at least one of these reacting molecules must divide or reduce, and therefore they cannot be considered simple substances, according to Lavoisier's definition. This conclusion allowed Avogadro to propose the formula $H2O$ for water, instead of HO, using stoichiometric experimental relationships and densities of these gases, measured in similar conditions of pressure and temperature, as well as to formulate ammonia as $NH3$, suggesting implicitly molecules of hydrogen, oxygen, and nitrogen as $H2$, $O2$, and $N2$, as formulations compatible with experimental data as well as with his hypothesis.

By using a clear algebraic form, we can demonstrate these same conclusions combining equations (1) and (3). Using water as an example, replacing Pat(H)/Pat(O) = d(H)/d(O) in equation (1), we consent to:

$$(m(H)/m(O)) / (d(H)/d(O)) = (n(H)/n(O))$$

This relationship allows us to calculate the relationship of the indexes of the formula for water:

$$(n(H)/n(O)) = 2$$

It is worth highlighting that this conclusion is obtained by using only experimental data—the densities of hydrogen and oxygen gases—and analytic measurements of the ratio in weight of hydrogen and oxygen to obtain water, as well as the hypothesis of Avogadro. Regarding the formulation of the molecules of H_2 and O_2, they are the result of obtaining the simplest integers for the reaction of the formation of water, keeping in mind that the decrease in volume found experimentally doesn't correspond to the reaction:

$$2H + O = H_2O$$

but, rather, to the reaction:

$$2H_2 + O_2 = 2\,H_2O$$

This fact establishes the concept of a homogeneous molecule: that is to say, constituted by two or more atoms of the same element, specifically the molecules H_2 and O_2 for gases hydrogen and oxygen at room temperature. A similar reasoning applies to the aforementioned case of the formation of ammonia, formulating the nitrogen molecule as N_2.

By clearly establishing atomic weights and their difference from molecular weights for hydrogen, oxygen, and nitrogen gases, a firm base was established for calculating atomic and molecular weights in general, completing an important stage of the solution to the chemical "puzzle." This stage of evolution of knowledge was arrived at when, in 1860, Cannizzaro, Italian like Avogadro, presented his chart of atomic weights at the first International Congress of Chemistry which was now appearing on a grand stage. The genius of Avogadro was not very well-known in his time, perhaps because he worked alone and his work was poorly disseminated, as well as for geographical reasons and perhaps a political distrust of science originating south of the Alps. The so-called Avogadro number (defined as the number of molecules or atoms contained in one gram-mol—that is to say, a quantity of matter whose mass expressed in grams is numerically equal to the molecular or atomic weight) was not measured until 1912, when X-ray diffraction data was used, allowing the measurement of interatomic distances in crystalline solids.

D. THE PERIODIC TABLE

Toward the mid-nineteenth century, the acquisition of knowledge in the chemical science was mainly focused on obtaining new "simple substances" of Lavoisier that already definitively were designated as elements or fundamental units of material reality, atoms. The measurements of the composition of compound substances, the molecules, allowed their formulation in connection with the elements, using as a unit the atomic weight of the lightest discovered element, hydrogen. Until then, differences persisted among measurements of the reported atomic weights, which could be attributed to experimental errors generated in obtaining elements and their purification procedures, as well as to imperfections of instruments used for measurement. The Avogadro hypothesis based on the mechanical properties of gases, such as their compressibility and density, allowed for the formulation of the water molecule as H_2O, which in turn was used as a base to calculate atomic weights, which became useful for the classification of the elements in a periodic table ordering them by ascending values.

This table, as its name indicates, shows physical and chemical periodical properties of the elements, suggesting a basic foundation of matter in units of lesser order than the known atoms. However, as a first approximation, the atom of hydrogen was considered the fundamental unit of matter. The variation of the density of elements in connection with their atomic weights, reported by Julius Lothar Meyer, and the table of the periodic classification of the elements of Dmitri Mendeleev, were published around the same time, in 1870 in Germany and Russia. The originality of this classification, especially the one presented in Table 6, like all great historical discoveries, continued to be revealed as new elements were isolated, corresponding to empty spaces that Mendeleev had assumed, anticipating some of their physical-chemical properties that were ratified experimentally.

It is worth noting that this remarkable classification of the elements introduced for the first time two fundamental concepts that characterize matter's diversity: atomic composition and an order to its chemical affinity properties. To understand fully the formidable task that Mendeleev carried out with his periodic table of the elements, let's return to the simile of the crossword or jigsaw puzzle. An ordered relationship was searched for using atomic weights of the elements and their properties that allowed for the establishment of groups or families of elements similar in their chemical reactivity. Objections and difficulties in this quest for experimental and theoretical order were found of the following tenor:

1. In connection with the numeric order of the atomic weights, it was believed that the hydrogen atom was the fundamental element that formed the other atoms; however, the fact that atomic weights found experimentally in this unitary base were not integers, challenged this hypothesis.
2. Not all the elements could be associated in clearly defined groups with similar characteristics.
3. It was suspected that many elements were yet to be discovered.

In any event, managing through these difficulties, Mendeleev classified the elements in two main categories:

1. The typical elements that correspond to the first fourteen known lightest elements, excluding hydrogen, and that clearly formed similar families (horizontal lines of two elements) that define periods of seven elements (columns)

2. The other known elements that include elements similar to the typical ones and others we now call transition elements, generally of metallic character, organized in periods of seventeen elements.

With these rules in mind the periodic table (Table 6) was proposed by Mendeleev, who ordered it in columns that showed periods of seven typical light elements that gave the key to obtaining, in rows, families that were chemically similar.

The three following periods were each comprised of seventeen elements, of which seven were typical elements

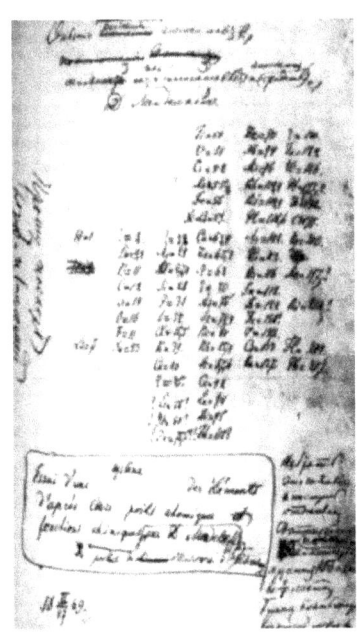

Table 6: MENDELEEV'S PERIODIC TABLE (Manuscript)

that belonged to the families found in the first section of light elements, and the other ten were transition elements whose physical-chemical likenesses were not understood so clearly.

			K = 39	Rb = 85	Cs = 133	"	"
			Ca = 40	Sr = 87	Ba = 137	"	"
			"	Yt = 88?	Di = 138?	Er = 178?	"
			Ti = 48?	Zr = 90	Ce = 140	La = 180?	Th = 231
			V = 51	Nb = 94	"	Fa = 182	"
			Cr = 52	Mo = 96	"	W = 184	Ur = 240
			Mn = 55	"	"	"	"
			Fe = 56	Ru = 104	"	Os = 195?	"
	Typical Elements		Co = 59	Rh = 104	"	Ir = 197	"
			Ni = 59	Pd = 106	"	Pt = 198?	"
H = 1	Li = 7	Na = 23	Cu = 63	Ag = 108	"	Au = 199?	"
	Be = 9,4	Mg = 24	Zn = 65	Cd = 112	"	Hg = 200	"
	B = 11	Al = 27,3	"	In = 113	"	Tl = 204	"
	C = 12	Si = 28	"	So = 118	"	Pb = 207	"
	N = 11	P = 31	As = 75	Sb = 122	"	Bi = 208	"
	O = 16	S = 32	Se = 78	Fe = 125?	"	"	"
	F = 19	Cl = 35,5	Br = 80	I = 127	"	"	"
Short Periods			Large Periods				

This periodic table naturally was perfected and completed over time as experimental data became available, until it was theoretically supported by means of the quantum theory of the atom, which, in a spectacular form, provides a mathematical model that replaces the atomic weights with quantum numbers that justify the periodic nature of the physical-chemical properties of the elements.

E. THE CHEMISTRY OF CARBON

The development of the chemical atomic theory was carried out with materials called minerals, which excluded the plants and animal life of whose composition the presence of the element carbon was only known as a general characteristic, observed by means of the combustion process. The analysis of simple compounds from the chemistry of carbon, such as the hydrocarbons (derived from petroleum), alcohols (spirit of the wine), and acids (spirit of the vinegar), gave place to the so-called organic chemistry, the name of which indicates its relationship to living organisms. This analysis paved the way for the study of life from a point of view of its material composition, relating it for the first time with the new science of the atom.

The formidable intuition of chemical pioneers allowed for the advancement in the theory of molecular structure given by the formulas of Lavoisier, called empiric, toward structural formulations that describe compound substances by means of molecular models of atoms drawn as spheres that are joined by bars or lines. These models give to the group as a whole a rigid structure characterized by a figure in three dimensions (see Figure 8).

These space models looked to explain the experimental fact called isomerism, which refers to the differences between compounds that have the same composition or empirical formula, but manifest very different physical and chemical properties. For example, consider hydrocarbons, with an empirical formula of $C_n H_n$, which indicates a series of compounds that have the same relative proportion of carbon and hydrogen, and different molecular weights; $C_2 H_2$ (gas acetylene) and $C_6 H_6$ (liquid benzene) correspond to very different space models, as shown in Figure 8.

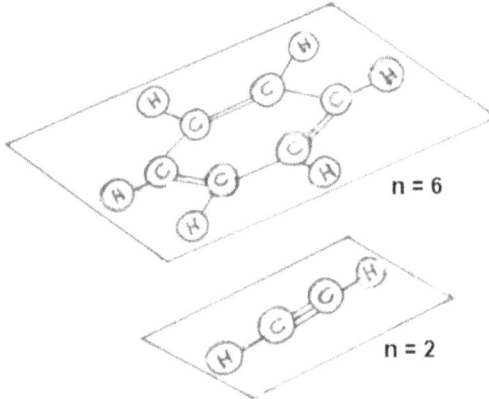

Figure 8: EMPIRICAL FORMULA Cn Hn; STOICHIOMETRIC ISOMERS

It is worth emphasizing the flat molecule of benzene ($C_6 H_6$), proposed by Friedrich August Kekulé, who sought to explain the chemical properties of the isomers derived from this hydrocarbon, which have great importance in the study of organic chemistry. The genius of the concept of chemical unions depicted spatially, as well as many other structural principles sensed intuitively by the scientific pioneers, have been corroborated by spectroscopic methods related to the quantum physical-chemistry theory that was developed decades later.

F. PHYSICAL CHEMISTRY

It is common for scientific pioneers to propose brilliant theories of a tentative nature that are the product of the creative intuition of philosophers and scientists. Karl Popper and John C. Eccles,[1] in their study of the mind and brain, refer to this human activity as "World 3," which eliminates the shades that intervene in the communication between World 2, the observer, and World 1, the physical, external, and objective world.

Newton and Galileo's celestial mechanics and the theory of the microcosms of Dalton and Lavoisier were integrated by means of the physical-mathematical workings that related them in logical form. The Maxwell-Boltzman kinetic theory of gases, using the formulation from classical mechanics, produced the formula for the ideal gases of Dalton:

$$P.V = nRT$$

introducing the constant R of the gases and the number of moles n which both imply the hypothesis of Avogadro. In turn, the famous phlogiston of the Greeks disappeared definitively from the scientific scene, with the equivalence of heat and mechanical energy or movement.

The atomic theory, intuitively created within the realm of chemistry, and its valuable contributions to the synthesis of the elements made possible its continuity in the physical study of the atom, by means of ingenious experimental methods called spectroscopy, which brought about the mathematical formulation of quantum physics.

5-6 QUANTUM THEORY

Just as the special theory of relativity led to a new understanding of space-time that revolutionized Newtonian mechanics, based on the experimental fact of the constant nature of the speed of light regardless of the coordinate system in which it is measured, quantum theory establishes the discontinuity of the physical magnitudes in the experimental environment of elementary particles as we measure the interaction of radiant energy with matter. The analysis of this theoretical-experimental process leads to a model for the atom and its constituents that implies a new understanding of nature. It moves away from our macroscopic world, very far even from the dimensional scale of the atom.

---

[1] Karl Popper and John C. Eccles, *The Self and Its Brain* (Routledge, 1984).

5-6-1 Origins and Their Development

A. PLANCK'S POSTULATE

In the historical development of atomic theory, the concept of quantification or discontinuity of energy appears as a mathematical hypothesis,
known as Planck's formula (1900):

$$\varepsilon = h \cdot v$$

where (ε) represents the changes or "energy jumps" that occur in the processes in which radiant energy interacts with matter, supposedly comprised
of atoms or fundamental discrete units. On the other hand, this energy is
related by means of the constant (h), with the frequency (v), from the
electromagnetic radiation of Maxwell's theory. This fundamental hypothesis has two revolutionary aspects in physics: the quantification of energy
or the affirmation of its natural discontinuity and its relationship with the
undulatory theory of electromagnetism. Its postulation was a theoretical
premise devised by Max Planck to derive a mathematical relationship of
the intensities of electromagnetic radiation of the "black body," as a function of the distribution of frequencies at determined temperatures. Figure
9 corresponds to such intensities of radiation of emission of the black
body, calculated as a function of the frequencies at a given temperature,
corresponding to the hypotheses of Planck (a) and the classical theory (b).

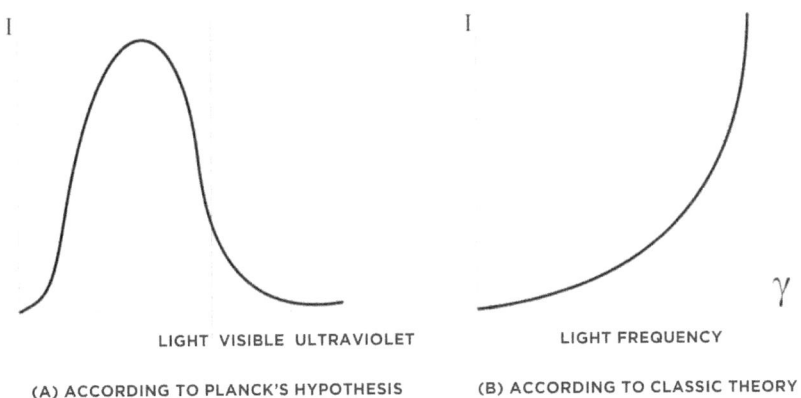

LIGHT VISIBLE ULTRAVIOLET   LIGHT FREQUENCY

(A) ACCORDING TO PLANCK'S HYPOTHESIS   (B) ACCORDING TO CLASSIC THEORY

Figure 9: SPECTRAL DISTRIBUTION VS. INTENSITY OF THE BLACK BODY
RADIATION

The prediction (b) from classic theory was called the ultraviolet catastrophe, which indicates that the assumption of continuity of the scale
of energy leads to a distribution of the radiant energy totally different

from the experimental data, especially in the area of high frequencies (ultraviolet); in contrast, Planck's equation predicts distribution (a) that agrees with the experimental data.

Two other phenomena secured the quantum hypothesis: the photo-electric phenomenon and the electromagnetic emission spectra from hydrogen, as modern atomic theory definitively took direction. We must consider Einstein's brilliant simplicity in his quantitative explanation of the photoelectric phenomenon (1905) that won him the Nobel Prize, as well as the genius idea of the atomic model of Rutherford and Bohr, which, with the quantified orbits of the electron that rotates around the small nucleus as the planets around the sun, gave the first explanation of the spectral discontinuity of electromagnetic radiation as it interacts with matter.

B. THE PHOTOELECTRIC EFFECT

In the photoelectric phenomenon, the electromagnetic radiation, light of frequency λ, impacts a metallic plate (cathode) that is inside a vacuum tube (see Figure 10). It thereby produces an electric current I (m.a) that closes in an external circuit through the anode, which captures inside the tube electrons emitted by the cathode by means of the radiation that it receives. Externally the tube electrodes are polarized, in opposite directions, with a variable potential V that can nullify the current in the circuit.

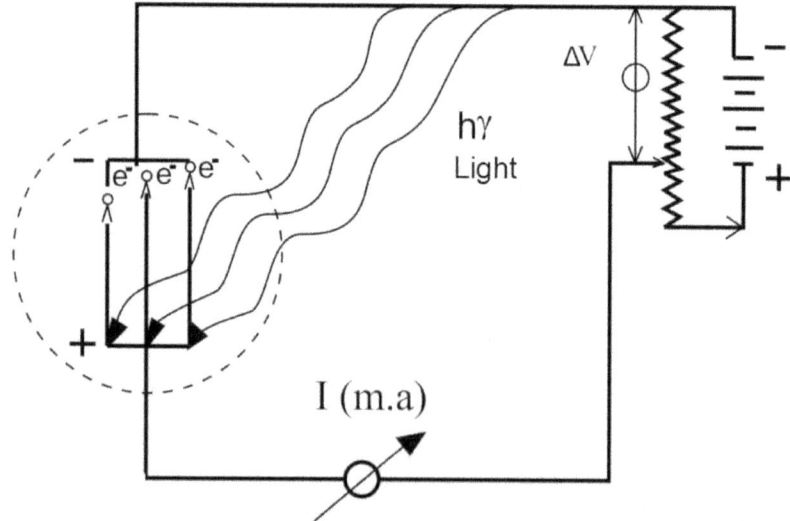

Figure 10: EXPERIMENTAL OBSERVATION OF THE PHOTOELECTRIC EFFECT

The experimental data obtained when varying the intensity and frequency of the incident light, the metal of the cathode, and the applied voltage can be summarized as:

1. For each type of metal used as a cathode, a frequency v_0 exists, below which current is not produced in the circuit, independent of the intensity of the incident light.

2. For frequencies $v > v_0$, a current is produced that can be suppressed, I (m.a) = 0, by means of the external potential ΔV.

Einstein interpreted the phenomenon by means of the equation:

$$h v = h v_0 + e.\Delta V$$

which can be represented by a direct line, $v = v_0 + e/h.\Delta V$, to calculate the threshold frequency v_0, the constant e/h, the relationship between the electronic charge, and the constant, h, of Planck's equation (see Figure 11).

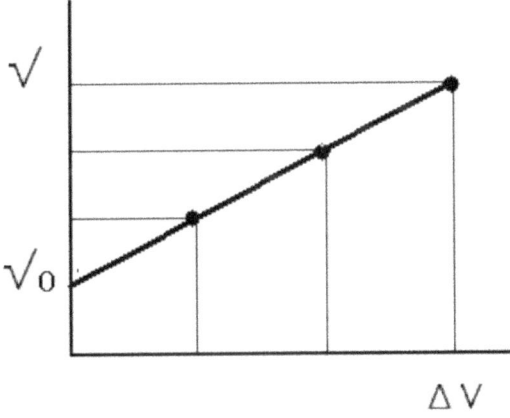

Figure 11: REPRESENTATION OF EINSTEIN'S EQUATION FOR THE PHOTOELECTRIC EMISSION

$$v = v_0 + e.\Delta V$$

The term (e.ΔV), where e is the charge of the electron, measured independently at that point in time by the group of Joseph John Thomson (1900–1905), is the required energy to stop the electrons emitted by the metallic plate of the cathode, with an energy $\frac{1}{2} m v^2 = e.\Delta V$, where m is the mass of the electron and v its speed. The term $h.v_0$ is the required energy to extract the electrons from the constituent atoms of the cathode, and therefore is unique for each element.

It is important to highlight the novel aspects of this experiment and its interpretation that depart from classic physics:

1. The radiant energy that is captured by the electrons individually depends on the frequency ν according to the hypothesis of Planck, and not on the intensity of the electromagnetic radiation, which, in classical physics, depends on the amplitude of the wave.

2. The interpretation of the phenomenon suggests that the electromagnetic wave behaves as a stream of particles that transports its energy in discrete packages and not in a continuous form. On the other hand, the total energy or intensity of the radiation depends on the number of particles (called photons) by unit of area that transfer their energy individually to the electrons, that is to say, one to one in a quantum form.

C. THE THOMSON-RUTHERFORD ATOM

The model for the hydrogen atom, the simplest of the elements, derives from the model created in 1898 by Thomson, who studied electrically charged particles and their trajectories when they are subjected to electric and magnetic fields. Initially it was thought that neutral atoms were comprised of negatively charged particles, called electrons, absorbed within positively charged particles that had a much bigger mass. The atoms, when losing electrons, called β rays, simultaneously generated positively charged particles, designated particles α, discovered in the spontaneous radioactive decay of radium by the Curie chemists. In 1911, Ernest Rutherford designed an experiment to measure the size of the atom using particles α while being dispersed by atoms. This important experiment allowed for the creation of an atomic model substantially different from the previous one, structured in the way of the solar system, with a small nucleus (charged positively), surrounded by electrons filling most of the atomic space.

The size of the atomic nucleus was calculated based on the dispersion of the particles α, endowed with high speed and a much bigger mass than the electrons, which collided with a thin sheet of gold.

Figure 12[1] shows schematically the device used: a radioactive substance generates the particles α (nuclei of helium) that are funneled in a stream that collides with the sheet of gold located in the center of a device called a

---

[1] Leonard B. Loeb, *Fundamentals of Electricity & Magnetism* (New York: John Wiley & Sons, 1947).

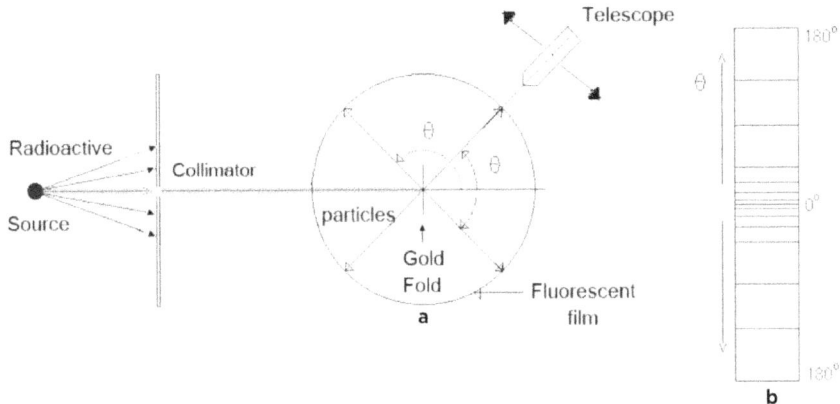

Figure 12: RUTHERFORD'S SPECTROMETER

spectrometer. This allows for the measurement of the angle (θ) of dispersion of the particle α when colliding with the gold atoms, as they are detected by a fluorescent film, by means of a telescope. The gauge at the right of Figure 12 shows the statistical distribution of the particles dispersed at different angles, indicating that most are deviated to low angles close to 0 degrees, and very few with angles larger than 45 degrees. The mathematical analysis of this experimental distribution was helpful to Rutherford in proposing his atomic model, which allowed for the calculation of the radius of the nucleus (10^{-13} cm) and its positive charge concentrated in a very small region in connection with the size of the atom (10^{-8} cm).

This model suggests a great "empty" space within the structure of matter, space where negatively charged particles would be located that justified the neutrality of the atom. Consequently, the investigative focus was centered on these particles, called electrons, which were studied by means of the generation of cathodic beams in vacuum tubes, mentioned in the photoelectric effect.

Initially the relationship (e/m), of charge (e) and the mass (m), was measured by means of the visualization of the trajectory of electrons that were submitted to forces or electromagnetic fields. Independently, the charge (e) was for the first time estimated by Robert S. Mulliken with a device that measured the displacement of small drops charged in an electric field; in this form an estimate of the value of the electronic mass was obtained that appeared to be much smaller than the atomic mass and consequently suggested that the nucleus not only concentrated the positive load but also its mass. The consolidation and improvement of the atomic model would need to include different stages that are presented today by three spectroscopies:

1. The interaction of particles, such as the electrons and particles α, with the matter (atoms) under energy conditions similar to those to which we have referred.

2. The interaction of electromagnetic radiation and the atom, by absorption or emission.

3. The interaction of elementary particles with the atomic nuclei under conditions of high energy produced by particle accelerators.

The first two provide us with the fundamental base to consolidate an atomic model consistent with the facts of physical chemistry and that generates the postulation of the quantum theory. With the third spectroscopy one studies elementary particles in general and their interactions, by generating particles that have acquired very high energies compared with the conditions that we normally find ourselves in our earthly environment, and that presumably occur at a cosmic level in celestial bodies such as the stars.

The first two spectroscopies played such a preponderant role in establishing quantum mechanics, that we consider it indispensable to present them in certain detail in order to achieve a better understanding of their postulates. The third spectroscopy generalizes quantum theory to the nuclear environment and consolidates the new physics called quantum electrodynamics.

D. EMISSION AND ABSORPTION SPECTRA OF GASES

The emergent atomic model from all these experiences pointed toward the concentration of the atomic mass in a small, electrically charged nucleus as positive, and electrons with a charge that occupied the "empty" space. How are those electrons located around the nucleus, in such a way to explain the atomic stability compatible with the dynamics of their energy states that define their interaction with electromagnetic waves?

The study of the emission of electromagnetic radiation of gases subjected to electric discharges gave rise to measurements of wavelengths of that radiation, by means of an apparatus called a spectrophotometer. This apparatus functions by means of a diffraction grid or a prism (see Figure 13) that unfolds geometrically in lines creating the image of a slit that receives the light coming from the sample subjected to experimentation. It was found that each one of the lines of this image, or emission spectrum, is unique for each element. It was also found that the absorption spectrum of a gas consists of a brilliant background crossed by dark lines that correspond to missing emission wavelengths. In the spectra of solar light and other celestial bodies were found characteristic lines of absorption spectra of elements, such as hydrogen and even new elements like helium, giving origin to its name, which means "sun."

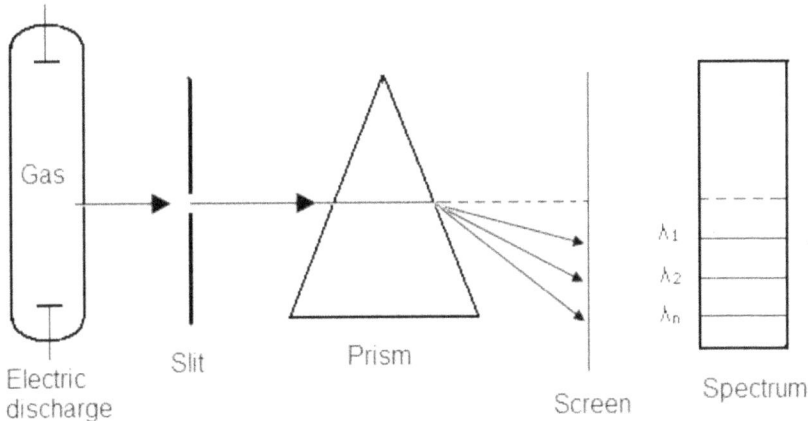

Figure 13: SPECTRA OF EMISSION OF THE ELEMENTS

The spectrum of the gas hydrogen has special interest because of its simplicity, the interpretation of which originated in Bohr's theory of the atom. Since 1885, Johann Jakob Balmer and others had found an empirical formula that orders the wavelengths (λ) mathematically for the spectrum of hydrogen, as follows:

$$1/\lambda = R \left(1/n_x^2 - 1/n^2 \right)$$

where R is the Rydberg constant, and n_x and n are integers that for the visible zone of this spectrum are $n_x = 2$ and $n = 3,4,5 \ldots$ (variable). Equally, other series of lines of the spectrum of hydrogen are obtained for values of $n_x = 1, 3$ for the ultraviolet and infrared zones, respectively.

E. BOHR'S ATOM

This model, for its simplicity, constitutes an important advance in understanding atomic structure: It allows for the calculation of spectral lines of hydrogen and the Rydberg constant R, utilizing classic mechanics and the hypothesis of the quantification of energy levels. At the same time, it relates atomic parameters, such as e,m,d, with the physical constants h and c, in a way that is consistent with many measurements of experimental variables that are presented in Table 7.

Bohr interpreted the lines of the spectrum of the H atom to correspond to transitions of the electron among circular orbits that have quantum levels of constant energy (see Figure 14). This hypothesis implies a stable atomic model, in disagreement with the classic electromagnetic theory, which establishes that an electric charge as an electron in movement irradiates electromagnetic waves in a continuous form, reducing its energy as it moves closer to the nucleus.

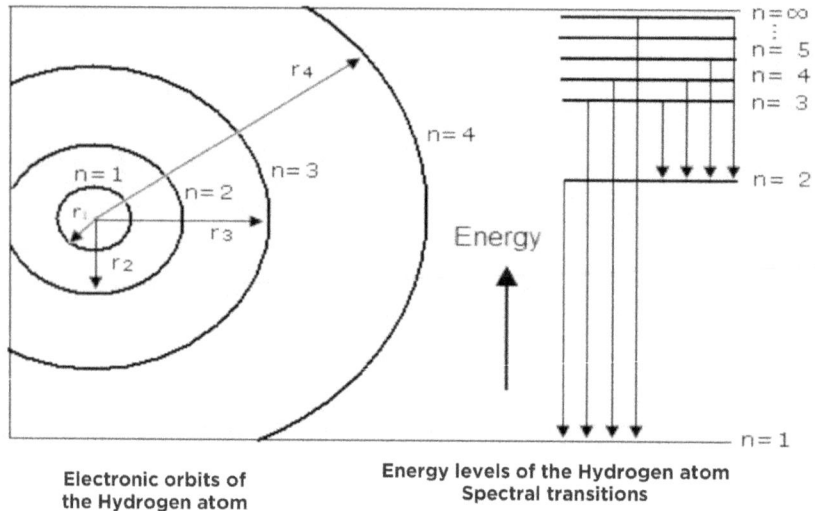

| Electronic orbits of the Hydrogen atom | Energy levels of the Hydrogen atom Spectral transitions |

Figure 14: BOHR'S ATOM

BOHR´S MODEL AND ATOMIC PARAMETERS

Bohr's model allows us to calculate the energy of the hydrogen atom for different orbits or stationary states, introducing a quantum condition similar to that of Planck, in connection with angular momentum, p, as follows:

$$p = (m.v) . r = n. h/2\pi$$

Here r is the radius of the orbit, m and v the mass and electronic speed, and n is an integer quantum number. The energy of the electron is calculated in a simple form, applying Coulomb's formula of attraction between the nucleus and the electron, in a similar way that Newton did with the planetary system. In such a way, the sum of the kinetic energy and potential is obtained as follows:

$$E(total) = R (1/n^2),$$

where $R = - (2\pi^2 m\ e^4/ch^3)$ cm^{-1}; $r_{n\ (Bohr's\ radius)} = n^2 (h^2 /4\pi^2\ e^2\ m)$ cm

Comparing these formulas and Balmer's empirical formula of spectral series allows us to calculate the Rydberg constant R.

In Table 7, a comparative summary of the atomic parameters is presented that characterizes the atomic model of Rutherford-Bohr and provides us with a global vision regarding its validity.

EXPERIMENTAL METHOD	BASIC FORMULATION	MEASURED MAGNITUDES	CALCULATED PARAMETERS					RESULTS
			e	m	h	d	l	
1. Cathodic rays	$2V/H^2.p^2 = e/m$	V,H,p	e/m	e/m	—	—	—	e,m,h
2. Photoelectric effect	$\Delta V/(n-n_o) = h/e$	ΔV, v	h/e	—	h/e	—	—	e,m,h
3. H Atom spectroscopy Bohr model	$1/\lambda = R(1/2-1/n^2)$ $R = (2\pi^2.m.e^4)/c.h^3$	λ (visible)	$m.e^4/h^3$	$m.e^4/h^3$	$m.e^4/h^3$	—	—	e,m,h
4. Light diffraction	$2d$ (grid) $Sen\theta =$ $n.\lambda$	θ, d (grid)	—	----	—	—	λ	λ (visible)
5. X-ray diffraction	$2d$ (atomic) $sen\theta = n.\lambda$	θ, angle	—	—	—	d/λ	d/λ	d (atomic distance)
6. Crystallography	r (density) = $n.M.N/d3$ N (Avogrado number)	r (crystal d.) M (molecular weight) n (n. molecules/ unit cell)	—	—	N/d^3	—		d
7. Electrolysis	F (Faraday) = N.e	F (electronic charge/ equivalent) e (electron charge)	Ne	—	—	—	—	N

TABLE 7: ATOMIC PARAMETERS

Table 7 presents seven experiences that allow for the measurement of some fundamental atomic parameters as follows:

The charge (e), the mass (m) of the electron, and the constant of Planck (h) are derived from the first three experiences. The fourth experience of the diffraction of light introduces the dimensions of longitude and time that are related directly to the macroscopic world. The diffraction grid (d) and the instruments used when measuring the speed of light (c) allow for the measurement of the wavelength (λ) and frequency ($v = c/\lambda$) of the luminous waves. On the other hand, experiences 5 and 6 present crystallographic measurements of different order, from the simple density of crystals, to the interpretation of X-ray diffraction that implies the geometric theory of classic crystallography and employment of the Bragg equation of diffraction. In this way we access the measurements of interatomic distances (d) and wavelength (λ) of

X-rays, using the number of Avogadro (N) and molecular weights (M) that go back to the basic concepts of physical chemistry (Section 5-5-4). The electrochemical experience of Faraday, among others, allows for calculation of the number (N), if one knows the electronic charge. The consistency of the atomic model of Thomson-Bohr is supplemented when comparing the radius of Bohr and interatomic distances (d) obtained from X-ray diffraction, which are of the same order. Nevertheless, the undeniable success of this model of the hydrogen atom presents difficulties when one tries to generalize to other atoms; on the other hand, it doesn't formulate a general theory that modifies classical mechanics and electrodynamics.

F. WAVES AND PARTICLES

James Clerk Maxwell's electromagnetic theory definitively established the wave nature of light. It shows properties such as diffraction or interference; nevertheless, it is found that its behavior in some experiences, as in the photoelectric effect, is explained by means of the corpuscular model—that is to say, as constituted by energy packages or corpuscles called photons. This modality is designated as the duality corpuscle-wave. Investigators such as Louis de Broglie wondered: If this duality could extend to particles in movement like electrons, whose corpuscular nature was given when measuring its mass, would it be possible to find experimentally that a beam of electrons would demonstrate diffraction phenomena?

De Broglie in 1923 proposed a theory of wave mechanics that describes "material waves"' associated with the displacement of free particles. What is the meaning of those waves associated with particles? An important aspect will be to find a mathematical formulation for those waves that can be related to classical mechanics. In this context quantum mechanics is born, a strange theory that oscillates between classic mechanics, including special relativity, and the volatility of particles that are simultaneously waves.

With the purpose of approaching a global understanding of the evolution of these theories, it is useful to go into certain detail of the mathematical expressions of the wave phenomenon that goes from mechanical and electromagnetic waves, to the material waves of de Broglie, giving rise to emergent quantum mechanics.

G. MECHANICAL WAVES

In mechanics, the theory of wave movement originally referred to a cyclical movement that spreads in a material medium, such as a string or the surface of the water of a pond. The disturbance originates at a given point, by a hand shaking the end of a string or a stone skimming the mirror of calm water, creating a wave that extends at a speed determined by the nature of the system, its origin, and the medium of propagation. Sound waves originate when vibrating objects, such as the surface of a drum or the arms of a pitch, extend their periodic movements into the air. We say that waves on a string or on the surface of water are transverse, because their movement has a perpendicular direction to its propagation speed; in contrast, sound waves are described as longitudinal waves, which means that the movement of the particles in the air has a parallel direction to its propagation speed (see Figure 15).

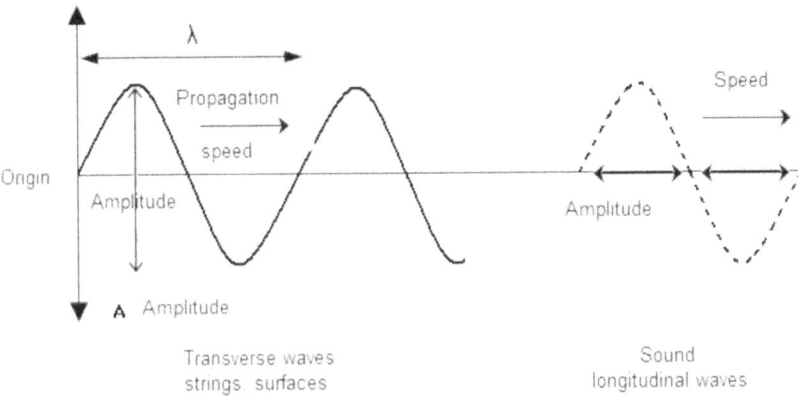

Figure 15: TRANSVERSE AND LONGITUDINAL WAVES

These movements are characterized by three magnitudes: the propagation speed, v, measured, for example, in meters per second; the frequency, ν, of vibration taken from the origin, measured in cycles per second; and the wavelength, λ, measured in meters. These variables are related as follows:

$$v = \lambda \cdot \nu$$

WAVE EQUATIONS AND MATHEMATICAL SOLUTIONS

Mechanical waves are visualized, in the world of our senses, by the periodic movement of a material medium that transmits the displacements of the origin. The mathematical general formulation, called wave equation, is expressed in one dimension, as follows:

$$\delta^2 \Psi(x,t)/\delta x^2 = (1/v^2) \ (\delta^2 \Psi(x,t)/\delta t^2) \qquad (5\text{-}6\text{-}1)$$

This equation has a general solution, where A is the maximum amplitude, as follows:

$$\Psi(x,t) = A\ \psi(x-vt)$$

A particular solution of the wave equation is expressed according to a trigonometrical function, sine or cosine as follows:

$$\Psi(x,t) = \text{sen}\ \lambda(x/\lambda - vt)$$

And in radian units as follows:

$$\Psi(x,t) = \text{sen} 2\pi\ (x/\lambda - vt) \qquad (5\text{-}6\text{-}2)$$

We can visualize this wave by giving it fixed values for the variables x, t; for example t = 0:

$$\Psi(x,0) = A\ \text{sen}\ (\theta); \quad \theta = (2\pi x/\lambda)$$

This represents an instantaneous picture of the wave at t = 0 that repeats itself for each value of x = nλ, or at values of θ = 2πn, for integer n, according to Figure 16, as follows:

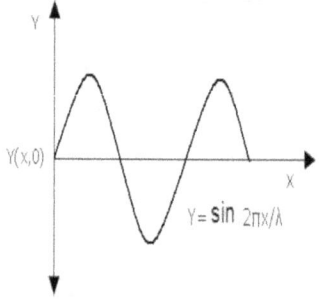

Figure 16: INSTANTANEOUS PICTURE OF THE WAVE AT T = 0

On the other hand, fixing a value for x, for example, x = 0 (origin), allows us to represent the function as follows:

$$\Psi(0,t) = A\ \text{sen}\ \omega\ t\ \ ;\ \omega = 2\pi v,$$

by means of the projection of the circular movement of a point with constant angular speed ω (see Figure 17):

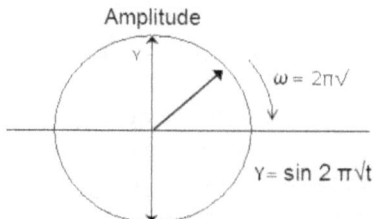

Figure 17: SIMPLE HARMONIC MOVEMENT A, X = 0

In natural wave phenomena it becomes necessary to consider overlapping waves that are combined to give a total effect. In what follows, some vectorial expressions that facilitate the summation of overlapping of waves are shown:

To facilitate calculations it is convenient to define trigonometrical expressions of waves, with parameters $\omega = 2\pi v$ and $\kappa = 2\pi / \lambda$, that denominate angular frequency (ω) and constant propagation (κ). Consequently, propagation speed is $v = \omega/\kappa$; in these terms the wave expression is:

$$y = A \operatorname{sen} (\kappa x - \omega t) \qquad (5\text{-}6\text{-}3)$$

It is worth reemphasizing that the three functions 1, 2, 3 express the periodicity of the wave in terms of space and time, in parametric form by the wavelength (λ) and by the frequency (v), or by the propagation constant (Đ) and the angular speed (ω).

A more general formula for waves, using sine and cosine functions, is the Moivre function that uses complex numbers, a + ib, and is expressed in exponential form:

$$r\,e^{i\theta} = r(\cos\theta + i \operatorname{sen}\theta) = r \exp(i\theta)$$

We can represent this function in a vectorial form with its two components, $x = r \cos\theta$ in the real axis, and $y = r \operatorname{sen}\theta$ in the imaginary axis. Adding vectors is accomplished by arithmetic summation of real and imaginary components, as shown in Figure 18.

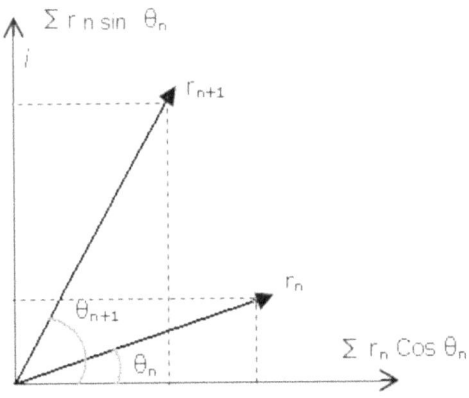

Figure 18: VECTORIAL SUMMATION OF WAVES

Adding waves is expressed by the summations of exponentials:

$$\Sigma_n r_n \exp(i\theta_n) = \Sigma_n r_n \cos\theta_n + i(\Sigma_n r_n \operatorname{sen}\theta_n)$$

Using wave equation (see Section 5-6-1), we will have for the sum of two waves:

$$A_1 \exp (i (\kappa_1 x - \omega_1 t) + A_2 \exp (i (\kappa_2 x - \omega_2 t)$$

H. ELECTROMAGNETIC WAVES

Classic electrodynamics studies forces that are exerted between electric charges. These are called electrostatic forces, if the particles are at rest, and magnetic forces if they are generated by electric charges in relative movement. In Maxwell's theory, radiation emitted by an oscillating electric charge (e) is described by means of an electromagnetic wave (see Figure 19), where the waves represent periodic variations of forces or electric fields (E) and magnetic fields (B) that spread at a speed (c). Figure 19 shows waves located in perpendicular planes that intersect in a line that indicates the direction of that speed.

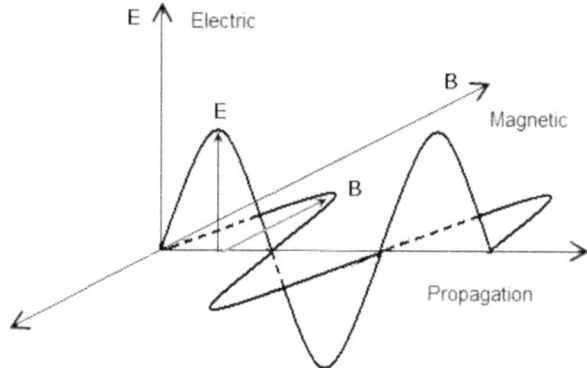

Figure 19: ELECTROMAGNETIC WAVE

Periodic variations of electromagnetic fields (forces) are expressed by functions E(x,t) and B(x,t) that mathematically have a similar form to mechanical waves. From Newtonian times, physical equations have been formulated using differential calculus, in which variations of magnitudes are expressed by means of derivatives with relationship to space (x) and time (t) (see Section 5-3). The wave equation in general is formulated according to equation 5-6-1, where derivatives are called partial because they refer to a single variable in each case, while keeping other variables constant.

Electromagnetic waves also follow the wave equation, where amplitudes refer to electric and magnetic fields independently:

$$\eth^2\,E,B(x.t)/\eth x^2 = (1/c^2)\,(\eth^2\,E,B(x.t)/\eth\,t^2)$$

These equations for electromagnetic waves in free space are deducible from Maxwell's equations of electromagnetic theory. According to quantum theory, these waves are associated with packages of energy or particles called photons; inversely we can expect that electrons or α-particles, which move at great speed in free space—that is to say, where there are no electromagnetic forces—also follow this wave equation.

I. INTERFERENCE, DISPERSION, AND INTERATOMIC DISTANCES

The phenomena called dispersion and diffraction of light, observed in nature as the rainbow, was studied in the laboratory with glass prisms and with artificial optic grids; on the other hand, classical physics elaborated the mathematical theory for mechanical waves that are observed in the movement of a string or in the interference originated in the surface of a calm lake when a stone is thrown that perturbs it. The similarity of these seemingly different phenomena was formulated generally by Maxwell, who proposed the electromagnetic oscillating nature of light, by means of a mathematical relationship called the wave equation that is common to all these phenomena.

Independent of the methods of the first spectroscopy, and to a certain extent of the second one, around 1912 measurement of interatomic distances was achieved in chemically simple compounds while in solid states, such as sodium chloride, or table salt, and some metals. To this purpose X-ray diffraction was used in crystals with atoms located at equal small distances ($d=10^{-8}$) from each other, forming a three-dimensional lattice that extends to the macroscopic level of the given crystal—that is to say, to a size (10^{-1} cm) that constitutes the monocrystal. In Figure 20(a), an ideal line of atoms is shown that diffracts X-rays that are electromagnetic waves with a wavelength of the order of 1 angstrom (10^{-8} cm); Figure 20(b) refers to diffraction of luminous electromagnetic waves that have a wavelength of the order of 10^{-5} cm, by means of a grating that is composed of slots located at regular distances (d), of the same order as the wavelength (λ) of the light used.

In both cases the diffraction phenomenon is present, which refers to the interference of such waves as the indicated ones, 1 and 2 in Figures 20(a) and (b), which originate in different places of the device that interferes with the incident ray. In Figure 20(a) the X-ray waves are "reflected" by two successive parallel planes of atoms of the crystal lattice (represented by lines), located at a distance (d), giving rise to the overlapping of waves 1 and 2 out of phase with each other by a distance (Δ). In a similar way, in Figure 20(b), the luminous waves that interfere originate in each one of the slots of the grid, located at a distance (d), previously established in their construction that defines geometrically, in both figures, the difference of travel of the two rays: Δ = 2d senθ. The grade of interference among the two waves will depend on the difference of travel between rays 1 and 2.

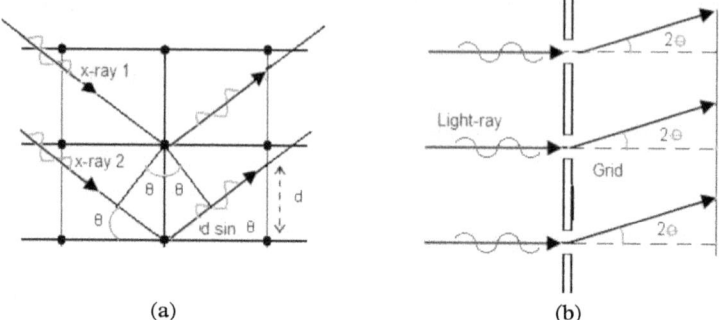

(a)	(b)
X-ray waves diffracted by a crystal lattice	Light waves diffracted by a grating

Figure 20: ELECTROMAGNETIC WAVE DIFFRACTION

On the other hand, in Figures 21(a) and (b) it is shown, in a general case, that waves 1 and 2 interfere. In Figure 21(a), it is observed that the waves are in phase when their maximum amplitudes coincide, thereby producing, when added, a maximum of diffracted intensity. In Figure 21(b), the two waves, with opposed phases, cancel each other out when the maximum amplitude of one wave coincides with the minimum of the other. These maximums occur when the traveling difference (Δ) of the diffracted rays 1 and 2 is made equal to an integer (n) for values of (λ). This condition allows us to relate the distance (d) and the length (λ), by measuring the angle (2θ) formed between the incident ray and the diffracted ray, by means of Bragg's equation:

$$2d \ sen \ (\theta) = n \ \lambda = \Delta$$

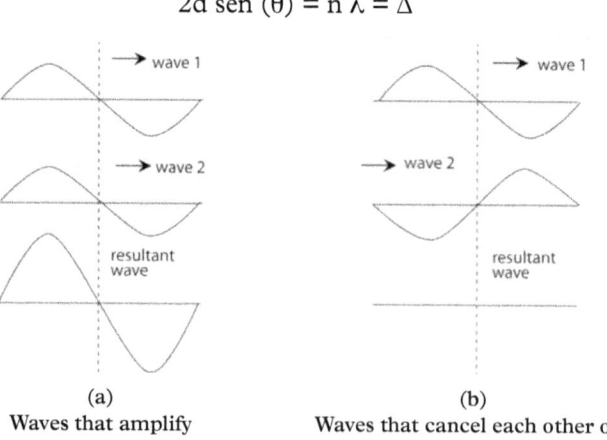

(a)	(b)
Waves that amplify	Waves that cancel each other out

Figure 21: INTERFERENCE OF TWO WAVES

In the experiment of diffraction of luminous waves through the grid, we measure the wavelength of light, comparing it with the width of the artificial slot, which belongs to the world of our senses. In the case of diffraction of X-rays through crystals, inversely we can determine the interatomic distances starting from the wavelength of the X-rays. However, in the first instance this experiment served to demonstrate that this radiation, originally of an unknown nature—giving origin to its name X—was definitively electromagnetic, in accordance with its polarization properties and its experimental origin, in which the kinetic energy of the electrons becomes radiant energy when colliding with a metallic plate. This phenomenon, when interpreted as the inverse of the photoelectric effect, suggested a wavelength of the order of 1 angstrom for X-rays, allowing evaluation by means of the Bragg formula of interatomic distances comparable with the ones obtained from the geometric crystallography theory.

Although the undulatory nature of moving particles, such as electrons, was not demonstrated experimentally until 1927 by means of the diffraction phenomenon, the material waves proposed by de Broglie constituted the fundamental ideas for the development of quantum mechanics by Schrödinger and Heisenberg.

J. WAVE MECHANICS AND THE PRINCIPLE OF UNCERTAINTY

In a first phase, de Broglie formulated the properties of wave functions associated with movements of free particles; in a development of greater breadth, Schrödinger formulated the general equation that constitutes the foundation of quantum mechanics.

Experimentally, in 1921 Arthur Compton demonstrated clearly that X-ray photons behave as particles when they collide with electrons. De Broglie was ahead theoretically in confirming the undulatory nature of free particles, when calculating their associated wavelengths by means of a heuristic demonstration—that is to say, plausible but not rigorous—along the following lines.

A photon with a mass m and a speed c has, according to the theory of relativity, a kinetic energy, $E = mc^2$, $E = (mc). c = p.c$; and by using Planck's equation, $E = hv = pc$; $h = p. (c/v)$:

$$h = p. \lambda$$

In a complementary way, in the Compton Effect, this important relationship allows us to calculate, by means of a rigorous demonstration, the wavelength of the photon dispersed when impacting an electron, as a function of the value λ_0 of the incident photon and of its angle ϕ of dispersion (see Figure 22).

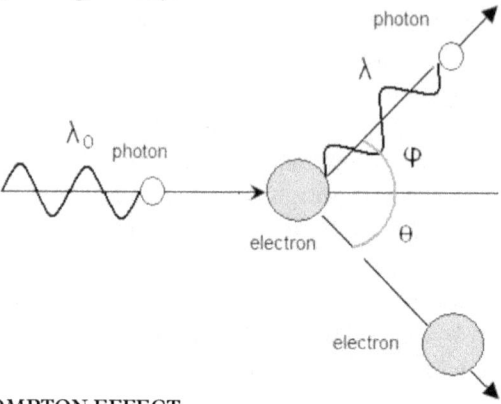

Figure 22: COMPTON EFFECT

On the other hand, it is important to highlight that this relationship between the wavelength and the linear momentum of a particle suggests that the nature of material waves has an implicit limitation on the precision of measuring the position of a particle (Δx) and of its momentum (Δp). When in an experiment we use a photon to determine the position of a particle, such as that of an electron, we alter its momentum. In this way, to specify the position of the electron, we should shorten the wavelength of the photon used, increasing the momentum of the electron at the same time and therefore varying in significant form the value of p of the electron. This restriction of the precision of these measurements (Δx, Δp) is expressed by the inequality:

$$\Delta x \ . \ \Delta p \ > \ h$$

This relationship, which at first sight refers to an incidental limitation of the precision of a measurement, constitutes a fundamental principle of the new quantum mechanics called the uncertainty principle, proposed by Werner Heisenberg. According to this "observational" position, the measurements of the physical magnitudes establish their own meaning that cannot be idealized by means of hypothetical procedures that ignore the limitations that nature imposes, such as with the values of physical constants c and h.

De Broglie, in his formulation of wave mechanics, looked for a mathematical function of the material wave associated with a free particle that expressed simultaneously its undulatory nature and the transitory characteristic in space that would describe its propagation. Pictorially, this wave function cannot simply be represented as a trigonometric function; it should be constituted by a pulse or a group of waves that travels at a certain propagation speed (see Figure 23).[1]

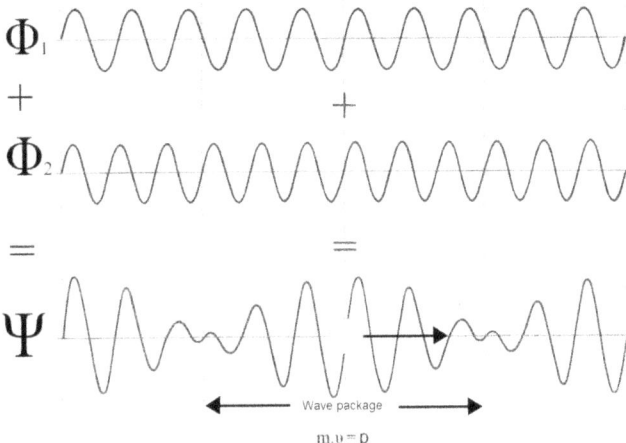

Figure 23: GENERATION OF THE WAVE FUNCTION OR DE BROGLIE'S WAVE PACKAGE ASSOCIATED WITH PARTICLE m THAT MOVES AT A VELOCITY Vg

WAVE PACKAGE AND THE UNCERTAINTY PRINCIPLE

In what follows, this "wave package" is expressed analytically, associated with a free-moving particle, which is consistent with the Heisenberg uncertainty principle.

In Figure 23, it is shown that the sum of two sine functions,

$$\phi1 = A \text{ sen } (kx-\omega t)$$
$$\phi2 = A \text{ sen } ((k+dk)x-(\omega+d\omega)t),$$

slightly displaced one from the other by values dk and dω—that is to say, by small differences of frequency and wavelength—generate a wave function,

$$\Psi = (\phi1 + \phi2) = 2A \text{ sen}(kx - \omega t).\cos(x.dk/2 - t.d\omega/2) \quad (5\text{-}6\text{-}4)$$

This wave function is constituted by wave groups, sine (kx - wt), that "travel" at a group speed, vg, produced by the modulation of a cosine wave that has constants km = dk/2 and ωm = dω/2. The speed of this encircling wave is vg

[1] Arthur Beiser, *Concepts of Modern Physics* (New York: McGraw-Hill, 1963).

$= d\omega/dk$. Its average wavelength is $\lambda m = 4\pi/dk$, and its frequency is $\upsilon m = \omega m$ $/2\pi$ (see formula 5-6-3).

This analysis, which can be generalized by means of exponential functions, suggests a model of the wave particle, characterized by a package or a group of waves that moves with a speed equal to the group speed of v_g, which implies a dislocation of the free particle that pictorially we can estimate as similar to the width of the group of waves:

$$\Delta x = 1/2\lambda m = 2\pi/\Delta k$$

On the other hand, this model of the waves of de Broglie drives us to a heuristic demonstration of the basic principle of quantum mechanics called the uncertainty principle, proposed by Heisenberg. The indetermination of the position of the particle (Δx) leads to the uncertainty of its speed v_g, and therefore of its linear momentum (Δp), which allows us to evaluate their interdependence.

The uncertainty Δp is given by the fundamental relationship $\lambda = h/p$, and by the definition of $k = 2\pi/\lambda$ that leads to the relationships:

$$k = (2\pi/h) \, p, \, \Delta p = (h/2\pi) \, \Delta k$$

The product of the indeterminations is:

$$\Delta x \cdot \Delta p = (2\pi/ \, \Delta k) \cdot (h/2\pi) \, \Delta k = h$$

The relationship $\Delta x \cdot \Delta p = h$ constitutes the uncertainty principle that relates the two indeterminations of position and linear momentum of a free particle, which is interpreted in its original Heisenberg form as the experimental impossibility of the simultaneous determination of the two variables. The determination or measurement of a variable such as the position of an electron in movement implies the use of another particle, such as a photon, that alters the linear momentum of the electron when colliding with it.

We can conclude that this principle of uncertainty is intimately bound with the special theory of relativity, the hypothesis of Planck, and the wave-particle model that are the foundations of quantum mechanics. Defined as a principle it is indemonstrable; its validity, like that of all scientific hypotheses, is based on its coherence with other theories. Stated another way, as scientific hypotheses may be fallible, they are considered valid so long as incompatible experimental facts are not found, or so long as they cannot be replaced by simpler, preferable hypotheses, according to the principle of Occam's Razor.

5-6-2 Establishment of Quantum Mechanics

The investigations of the physical nature of the structure of the atom, postulated by chemistry as the fundamental hypothesis that explains the diversity of matter and its reactions, showed that the emergent atomic model demanded fundamental modifications of Newtonian mechanics and classic electrodynamics. The discovery of the undulating nature of elementary particles that comprise the atom, as well as the corpuscular characteristics of electromagnetic waves, such as light, when interacting with matter, pointed toward a new interpretation of natural phenomena when analyzed directly at a very small scale, in comparison to accessible physical magnitudes in the environment of human senses.

Specifically, wave mechanics proposed wave functions associated with the particles in movement that suggested that the wave-particle duality should be formulated to the way of classical mechanics, with a differential equation that expressed the relationship between energy, space, and time by means of a function that also satisfied the fundamental equation for undulatory movement.

A. THE SCHRÖDINGER EQUATION

Great scientific transformations arise from experimental sets of data and from tentative theories that are summed up generally in hypotheses or fundamental principles, which in physics are expressed mathematically. Quantum mechanics was formulated simultaneously in different mathematical forms.

Erwin Schrödinger used linear differential equations, presented in their traditional form, while Heisenberg used algebraic relationships of matrices. These complementary procedures proved to be equivalent.

In a way that we can call heuristic we can understand the origin of the Schrödinger equation by means of an explanation that is close to the possible mental mechanism that led to its definition. The Greek term *heuristic* is used to designate an exploratory system in which feedback is used to obtain a consistency between two concepts or theories, which in this case refer to the theory of classic physics and the developing theory of quantum mechanics. In consequence, this methodology doesn't constitute a logical deduction of a theory such as quantum mechanics; however, it reminds us clearly that the principles of science, as important as they may seem, have value as hypotheses to the way of Euclid's postulates.

Einstein's relativity theory is considered to be classical physics; it is included in classical mechanics by way of the special theory, following heuristics, which has been denominated the principle of correspondence. In a similar way, according to this principle, quantum physics, which reigns in the environment of the microcosm, should be valid when being extrapolated to the world of our senses, without finding any inconsistency. It doesn't happen this way with gravitation from general relativity, which remains today like a monument of rationality that doesn't accommodate completely with quantum heuristics.

The fundamental equation of quantum mechanics, proposed by Schrödinger for the free particle, is:

$$(-\hbar^2/2m \; d^2/dx^2) \; \psi(x,t) = i\,\hbar \; d\psi(x,t)/dt \qquad (1)$$

$$\hbar = h/2\pi,$$

where $\psi(x,t)$ is the so-called wave function of x and t of the particle of mass m, in movement. It unites two quantum aspects: the particle's undulating nature, which expresses dislocation of the particle in space-time, and its relationship with classical mechanics, which allows the calculation of its energy states (E) and linear momentum (p_x) in the Hamilton form of $H = p^2/2m$ for the free particle, or, adding the potential V to which the particle is subjected, $H = p^2/2m + V$.

In the way of Richard Feynman,[1] it is fit to ask how Schrödinger arrived at this equation. Where did it come from? The answer is nowhere. There is no way of deducing it; it came from the mind of Schrödinger, in his quest to reach an understanding of the observations of the real world. However, complementary to the strictly objective attitude of most physicists, an analysis of the equation's historical development allows us to situate ourselves in a theoretical-experimental position that conjugates the two facets that imply the dual wave particle, satisfying our desire to inquire into metaphysics.

It was desired to employ the mathematical machinery of classical mechanics, which determines the magnitudes of energy and momentum of particles that interact with each other as a function of defined trajectories, in a form compatible with the strange description of wave mechanics, where moving particles are described by wave functions that don't determine trajectories to the style of Newtonian and Hamiltonian mechanics.

---

[1] Feynman et al., *Lectures on Physics*.

SCHRÖDINGER EQUATION AND THE WAVE FUNCTION

In this heuristic way, we can simulate the mind of Schrödinger, transporting ourselves to the scientific atmosphere of 1923. We are looking for a wave function $\psi(x,t)$ that satisfies simultaneously the two equations for material waves:

$$\delta^2 \Psi(x,t)/\delta x^2 = (1/v^2)(\delta^2 \Psi(x,t)/\delta t^2) \qquad (2)$$

$$H_{op} \Psi(x,t) = (p^2/2m)_{op} \Psi(x,t) = E \Psi(x.t) \qquad (3)$$

In these equations, v is the speed of propagation of the wave, E is the energy of the system, H_{op} is the Hamilton operator op of the energy, and p_{op} refers to the linear momentum operator. This equation introduces the concept of an operator that uses a symbol, op, which symbolizes an algebraic operation or of any other order; energy E corresponds to the Hamiltonian operator H (see Section 5-2) that is expressed as a function of operator p_{op} of the linear momentum. The wave function of de Broglie has an exponential expression:

$$\Psi(x,t) = A\, e^{-i(kx-\omega t)} \qquad (4)$$

equivalent to the trigonometric formula (see section 5-6-4) that satisfies the wave equation (2), with a group speed:

$$v_g = d\omega/dk = \omega_m/k_m = E_m/p_m$$

where subindex m refers to the average values of the particle.

If we replace v = E/p in the wave equation (2), we will have:

$$d^2 \Psi(x,t)/dx^2 = (p^2/E^2)\, d^2 \Psi(x,t)/dt^2$$

and differentiating function (4):

$$d^2 \Psi(x,t)/dt^2 = -\omega^2 \Psi(x,t) = -(E^2/\hbar^2)\Psi(x,t)$$

where $\hbar = h/2\pi$. Consequently we consent to:

$$d^2 \Psi(x,t)/dx^2 = -(p^2/h^2)\Psi(x,t)$$

or, reordering when multiplying by (1/2m):

$$(-\hbar^2/2m)\, d^2 \Psi(x,t)/dx^2 = (p^2/2m)\Psi(x,t) \qquad (5)$$

This last equation suggests its identity with the equation of Hamilton (3) and defines the energy and linear momentum operators:

$$H_{op} = (p^2/2m)\, op = (-\hbar^2/2m)(d^2/dx^2):$$

$$p_{op} = -i\hbar\, d/dx$$

Thus, we can express equation (5) in a condensed form:

$$\text{Hop } \Psi(x,t) = E \cdot \Psi(x,t) \qquad (6)$$

On the other hand, equation (5) is equivalent to the Schrödinger equation (1); the first members are identical and the second ones are equivalent when introducing function (4) of de Broglie:

$$i\hbar \, (d/dt) \; A \, e^{-i(kx-\omega t)} = \hbar \, \omega \, \Psi(x,t) = E.\Psi(x,t) = p^2/m. \; \Psi(x,t)$$

B. QUANTUM OPERATORS

The heuristic methodology of the correspondence principle[1] shows how the wave function of de Broglie, which satisfies the general wave equation, allows for the calculation of the energy of the particle by means of algebraic definitions of physical magnitudes from classical mechanics, by means of "operators" that act on the wave functions of particles and define differential equations in the new mechanics that simultaneously understand the classical expressions and the undulatory ones. These expressions are equivalent to the Schrödinger equation (1).

Thus with the definition of the operator p of the linear momentum, we deduce the Hamiltonian operator that allows us to formulate the kinetic energy of the particle. This mechanism, described for the operator p, can be generalized for any other physical magnitude, thereby establishing some rules to deduce the quantum operators, starting with the classical expressions of the physical magnitudes from classical mechanics.

The operators would be expressed as follows:

Physical Magnitude	Classical Mechanics	Quantum Mechanics
Linear Momentum	$p = m \cdot v$	$(p)op = -i \cdot \hbar \cdot d/dx$
Kinetic Energy	$E = p^2/2m$	$(H)op = - \hbar^2/2m.d^2/dx^2$

The physical magnitudes are calculated by means of equations that are expressed in general form:

$$(O)_{op} \, \Psi(x) = o \cdot \Psi(x) \qquad (6)$$

where o is the so-called proper value of the physical magnitude O, corresponding to the operator (op) O. In the case of the energy of a particle, we will have:

$$(H)_{op} \, \Psi(x) = E \cdot \Psi(x) \qquad (7)$$

which is the Schrödinger equation for the case of a system defined by the wave function $\Psi(x)$, of a stationary system, or independent of time.

---

[1] Jay Martin Anderson, *Mathematics for Quantum Chemistry* (W.A. Benjamin, Inc., 1966).

C. QUANTUM MODEL OF THE HYDROGEN ATOM; QUANTUM CHEMISTRY

When applying the Schrödinger equation to the electron of the hydrogen atom, which is subjected to a central electric field produced by the nucleus, we deduce the stationary quantum states characterized by wave functions $\Psi(x,y,z)$, and quantum numbers n,l,m that define its energy (E) and its form. The quantum number n, defined as the principal number, coincides with the number of the Bohr model; l and m define its directional characteristics that are related to the magnetic properties.

The wave functions of the hydrogen atom, even though they are not properly the same ones as those of the other atoms, serve as a base to delineate their periodic properties. Thus, it is possible to describe the electronic configurations for all the atoms, in a coherent way with the periods of the Mendeleev periodic table, proposed empirically in the beginning of the atomic theory.

The impact of this quantum model of the atom, in the development of chemical theory of the molecular structure, has allowed for the establishment in a physical-mathematical fashion of chemical concepts sensed by the chemical pioneers, such as bonding types between atoms, stereochemical properties, directed valence, and reactivity.[1] In particular, molecular spectroscopic properties have been related with their spatial structures, which are defined by means of symmetry relationships that are common to the stereochemical model and the corresponding molecular-wave function. This theory constitutes the mathematical foundation of chemistry called quantum chemistry. Numerous theoretical-experimental studies have served as bases for these important applications of the quantum theory, such as the ones developed in the dissertations of the author of this book.[2]

On the other hand, one can state that the success of the new quantum mechanics, which studied the hydrogen atom, created a firm base for the experimental and theoretical development in physics of elementary particles, generalizing its concepts to an n-dimensional space that imposed itself more and more as stranger properties of the subatomic world were discovered.

---

[1] Linus Pauling and E. Bright Wilson, Jr., *Introduction to Quantum Mechanics* (New York: McGraw-Hill, 1935).

[2] Jaime Pradilla-Sorzano, "Spectroscopic Studies of 'Octahedral' Copper (II) Complexes." (Ph.D. Thesis, Case Western Reserve University, 1972).

———, "Infrared Spectra and Normal Coordinate of Pt(II) Complexes," (M.Sc. Thesis, Case Institute of Technology, 1966).

Jaime Pradilla-Sorzano and John P. Fackler, Jr., *Inorg. Chem.*, 13, 38 (1974), 12, 1.182 (1973), 12, 1.174 (1973).

———, *Journal of Molecular Spectroscopy*, 22 (1967): 80–98.

Pradilla-Sorzano and John P. Fackler Jr., et al., *Inorg. Chem.* 18, 3519 (1979).

D. THE ELECTRONIC SPIN AND THE SUBATOMIC WORLD

The first property observed that was attributable "implicitly" to a subatomic particle was designated as the electronic spin. In the spectroscopic studies of emission or absorption of electromagnetic waves of hydrogen, the fine unfolding of some lines was observed that could not be explained by the quantum model of the Schrödinger equation. However, this quantum model allows for the calculation of magnetic fields produced by the orbital movement of the electron, which produces unfolding lines when subjecting the hydrogen sample to an external magnetic field. These facts suggested that an intrinsic property of the electron, called spin, generates a magnetic field quantified spatially in two opposite directions that classically could be visualized by the movement of the electron on itself. The coupling of the intrinsic magnetic field of the electron and its orbital movement gives place to the fine unfolding of the spectral lines.

It is well known that operators for the angular momentum of the electron, Op (L) and its components, deduced from their classic expressions, allow for the calculation of their corresponding proper values as a function of their quantum numbers, l, m, of the H atom, and therefore calculation of the energy corresponding to the Zeeman effect, which refers to the effect of an external magnetic field on the orbital magnetic moments of the electron. Therefore, it is plausible to assume that the unknown spin operators that refer to the intrinsic angular momentum of the particle should have algebraic properties similar to the operators of the angular momentums. It is very interesting to refer to the demonstration that obtains the proper values s, m_s of the spin operator, op (S), when considering that its operative properties are the same ones of the operators of the angular momentum L. This rigorous demonstration[1, 2] leads us to the following equations:

$$(S^2)_{op} \, \alpha = \, s(s + 1) \, \hbar^2 \, \alpha \qquad (S_z)_{op}\alpha = m_s \, \hbar \, \alpha$$

$$(S^2)_{op} \, \beta = \, s(s + 1) \, \hbar^2 \, \beta \qquad (S_z)_{op}\beta = -m_s \, \hbar \, \beta$$

In these equations, the $(S^2)_{op}$ refers to the scalar value or length of the spin vector S, and the $(S_z)_{op}$ to its component in the direction z, with proper values possible for quantum numbers s = 0,1,2 ... , o, s = 1/2,3/2...

---

[1] Henry Eyring, John Walter, and George E. Kimball, *Quantum Chemistry* (New York: John Wiley & Sons, 1965).
[2] Frank L. Pilar, *Elementary Quantum Chemistry* (New York: McGraw-Hill, 1968).

y $m_{s=+,-}$ 0,1,2,... $m_{s=+,-}$ ½,3/2,... that refer to any subatomic particle whose wave functions are α, β

The fine structure of some spectral lines for hydrogen allows for the measurement of energy, corresponding to the electronic transition between those states, and therefore to calculation of the spin value, s = 1/2, for the electron, when the interaction of its magnetic moment and the magnetic field that it is subjected to is considered. In a similar fashion it has been found experimentally that other particles, such as the atomic nuclei, are characterized by values of spin that are integers, or wholes of ½, like the theory predicts. The spin concept is generalized for any subatomic particle such as the proton, giving place to an important branch of spectroscopy called magnetic resonance.

Paul Dirac developed a general theory for the action of an electromagnetic field on an electrically charged particle, based on the interaction of a photon and an electron that incorporated the effects of special relativity. Of great impact was the theory's prediction that electrons can exist in two energy states, corresponding to the two positive and negative roots of the relativistic equation:

$$E = +,-((m_0 c^2)^2 + p^2 c^2)^{1/2}$$

where m_0 is the mass of the electron at rest. This theory was confirmed experimentally with the discovery of the electron with positive charge called the antiparticle of the electron or positron. Likewise, this property is generalized for all elementary particles that have the same mass, opposite charges, and alignment and counter alignment between their spin and magnetic momentum. On the other hand, this theory predicts the annihilation of a particle and its corresponding antiparticle, which dissipate into radiant energy. All these aspects are of great relevance to the theories of the origin of the universe.

We're led to wonder if these spin functions refer to dimensions outside of space-time. Initially, quantum scientists spoke of nonspatial dimensions or coordinates; however, as new subatomic particles were discovered and attempts were made to classify them, it became necessary to introduce new physical magnitudes that characterized them. To study the atomic nucleus, in the hope of finding its constituents and the forces that explained its stability, particles α were used, obtained from radioactive samples that demonstrated that the emergent radiations from these collisions had great penetrating power, with a mass close to the proton, without electric charge. Initially these nuclear particles, eventually called neutrons, were the basis for proposing the constitution of the atomic nuclei, based on their charge given by the number of protons, and its total mass, based on the number of protons and neutrons, which also explained the isotopes of each element.

With the development of particle accelerators, protons or nuclei of H were used, with high energy able to dismember the atomic nuclei. As higher energies were used, new particles appeared that were designated by Greek letters, until the alphabet was drained. Soon it became evident that the number of particles was not defined; they were not preexistent in the nucleus, as were protons and neutrons, and their characteristics had to be framed by the dynamics of their interactions. In this form, electrodynamics, which referred mainly to the interaction between electrons and photons, was expanded into the theory of chromodynamics, to encompass all the subatomic particles. New quantum rules of interaction were devised that were designated with such properties as the color, the strangeness, and flavors whose main actors are particles, called "quarks," that are grouped into two classes and have fractional electric charges. Thus, for example, particles as familiar as the proton and the neutron are constituted by three quarks in different combinations of their different "flavors," d and u, which explain transformations like the decomposition β of the natural radioactivity. Are these the particles that so far have not been detected experimentally: the new Greek indivisible atoms, without parts?

A new theory, called string theory, reduces quarks to strictly mathematical components, consequently belonging to the World 3 of Karl Popper, and—why not?—to the ideas of Plato. What, then, is left of the materialistic world of pseudo-science?

E. MEANING OF THE WAVE FUNCTION

This whole extensive quantum theory can be developed without assigning to the wave function a meaning in itself, instead considering it simply a kind of "mathematical wild card" that is useful for carrying out important calculations. The wave function is in general a complex number that can only be interpreted in a special "space," called Hilbert space. However, its square is a real number that has always been considered, from the initiation of quantum studies, to be associated with the probability of finding a particle in a certain position in space, perhaps because of the analogy with classic mechanics that relates the square of the amplitude of the wave to its intensity.

From the time of Niels Bohr until today, many have discussed the meaning of this function and its interpretation, which naturally involves quantum theory in and of itself. The reason for this controversy originates in the perplexity awakened by the experimental facts of the environment of the microcosm, as they relate to our habitual world from which classical theories are derived. From the time of Planck, physicists have tried to accommodate the majestic work of mathematical physics, which has had so many successes in science and technology, to the surprising events discovered in the environment of the atom. The quantum theory arose almost like an imposition from nature that threatened to knock down the fundamental concepts of science, based until then in a seemingly impeccable theory.

In the philosophical field, serious difficulties are also encountered; questions arise regarding new principles of the new science, such as the principle of uncertainty that implies that the measurement of the physical magnitudes, such as the position and speed of a particle, depend on the observer.

If we try to measure the position of a particle with accuracy, it's at the expense of the indetermination of its speed. Consequently, can facts exist independent of the observer? The external world depends on us.

If the theory demands that a particle pass simultaneously by several places and at the same time interfere with itself, for the purpose of explaining the diffraction of a photon or an electron in a couple of neighboring slots or in a grid, is this reasonable? Frequently, the mathematical rules that are necessarily assumed in quantum theory don't satisfy from the point of view of our daily experiences.

The new quantum theory is based on the classic one, modifying it, but at the same time it is not independent of its concepts.

To affront these conceptual conundrums, several views have been presented among scientists that have been broadly publicized. Among these interpretations, certain ones stand out, such as the original one by Bohr, or the Copenhagen interpretation, the interpretation of multiple universes,[1] and the interpretation of Bohm[2]; however, the basic mathematical formulation remains strong.

E-1. BASIC MATHEMATICAL FORMULATION

Physicists like Richard Feynman[3] assume a positivist attitude, presenting the validity of quantum theory as it relates to its capacity to predict observations correctly. These principles can be summarized as follows:

---

[1] David Deutsch, *The Fabric of Reality: The Science of Parallel Universes and Its Implications* (New York: Penguin Books, 1997).

[2] David Bohm and F. David Peat, *Science, Order and Creativity* (New York: Bantam Books, 1987).

[3] Richard P. Feynman, *QED: The Strange Theory of Light and Matter* (Princeton, NJ: Princeton University Press, 1985).

The objective of science is to predict events in relation to others. The dynamics used to deduce these predictions are called theories, which, in physics, are expressed by means of mathematical, numerical procedures.

In quantum mechanics we relate events or happenings by means of the wave function (Ψ), which designates an amplitude by means of a complex number $\Psi = A + iB$, which squared means the probability (P) that the event will occur with probability $P = \Psi^2 = \Psi^*\Psi$, where $\Psi^* = A - iB$ is the complex conjugate of Ψ.

The probability that an event will occur Ψ_T can be calculated in relation to other events $\Psi_1\ \Psi_2$........Ψi that determine it, by means of quantum rules to calculate the total amplitude Ψ_T in function of the functions Ψi.

In general two cases occur:
    a. events independent of each other
    b. successive events

In each case we have:
    a. $\Psi_{T=}\ \Psi_1 + \Psi_{2+}$........$\Psi$i
    b. $\Psi_{T=}\ \Psi_1 \times \Psi_2\ x$........$\Psi$i

These rules don't have any explanation; their validity lies in their capacity to predict nature by means of observation.

The observational position implies that to define an experimental fact, an event, it is necessary to define how it is observed; we should specify an experience.

E-2. INTERFERENCE OF PARTICLES IN TWO SLITS: A VIRTUAL EXPERIMENT

The diffraction effect or interference of particles from light, called photons, or from any other source, such as electrons whose trajectory, for example, is limited by means of slits or by other equivalent means, is usually presented to analyze these quantum rules.

According to the most modern interpretations of quantum mechanics, interference happens when, within the analyzed facts, there are indistinguishable alternatives, such as possible trajectories for a particle. In Figure 24, the experiment of the two slits is presented, which limits the trajectories of elementary particles, such as electrons or photons, described as energy packages and characterized by a single frequency. In this figure, the trajectories of the particles are presented schematically, leaving from a monochromatic source f, and having two options or trajectories when passing through slits a and b, which then impact the screen, giving place to the phenomenon of diffraction or interference that is characterized by producing on the screen or film fringes of light and shade designated by the letters c; d,d'; e,e'. In contrast, when the ray of photons has only one option of passing through a circular hole, it gives place to a circular image with high intensity in the center and dimness in its periphery, without presence of the diffraction phenomenon.

The phenomenon presented is interpreted as an interference of each photon with itself, having two possibilities for its choosing: either passing through slot a or b, to arrive at point c, or to the others d,e.....

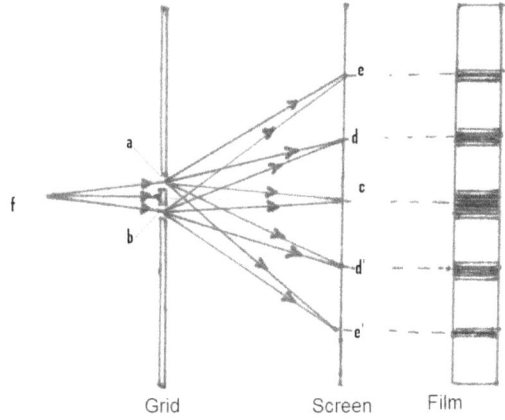

Figure 24: DIFFRACTION OF PARTICLES IN SLITS A AND B

Each particle is characterized by a wave function Ψ, whose amplitude can acquire different values according to the path the particle chooses. Thus, we designate different wave functions according to the trajectory that the particle follows:

$$\Psi\,(c, a, f\,), \quad \Psi\,(c, b, f\,)$$

$$\Psi\,(d, a, f\,), \quad \Psi\,(d, b, f\,)$$

The letters indicate the trajectory, beginning their sequence from the point on the screen backward, following the trajectory of the particle. Each pair (row) of values Ψ corresponds to a point on the screen, for example c, where a photon or particle impacts two paths, slots a and b. The interference occurs between this pair as follows:

$$\Psi(c,a,f) \;+\; \Psi(c,b,f)$$

This is abbreviated as follows:

$$\Psi_{\text{interference}} = \Psi(a) + \Psi(b)$$

This summation function represents the interference function of the photon (particle) with itself, when passing through two slots. This phenomenon can be generalized for n possible trajectories, according to the experiment considered. Feynman, in his book *Quantum Electrodynamics (QED)*, presents this analysis for a series of luminous phenomena including reflection, refraction, and diffraction, explaining characteristics of geometrical optics as the action of lenses, prisms, thin films, and so forth, in a way that is novel and brilliant for its simplicity and rigor. Feynman presents pictorially the mathematical way these vectorial summations are performed by adding arrows (vectors) that vary their direction according to the trajectory of each photon that is considered.

These summations of the values of the wave function for each one of the possible trajectories of the photon produces a resultant that represents the wave function Ψ_T, indicating the geometric rule that governs the phenomenon analyzed, such as reflection, which is characterized by the incident angle equal to its reflection or the inflection of the ray of light when passing from one transparent medium to another. This same analysis can be carried out in a more mathematically rigorous way by means of variable complex numbers (real number (A) + imaginary (iB)) that the wave function $\Psi(x,t)$ gives for each point (x) of the trajectory of the photon and that are represented in the Argand plane perpendicular to the speed of the particle (see Figure 25).

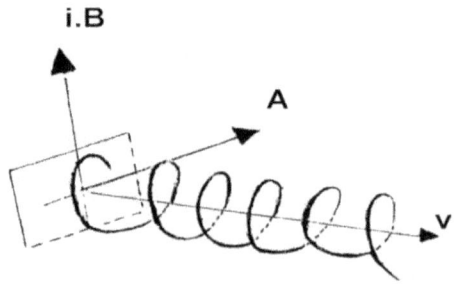

Figure 25: PHOTON WAVE FUNCTION REPRESENTED IN THE ARGAND PLANE

It is worth reflecting on the fact that this figure is a representation of the wave that should be differentiated clearly from the representation of a mechanical or electromagnetic wave (as in Figure 19). The line in helix form represents here the "trajectory" that the amplitude vector draws in the Argand plane, as the particle advances in its path, allowing for the addition of its components A and B arithmetically to obtain the resultant. The importance of this analysis of the evolution of the wave function resides in the meaning this function acquires in interpreting undulatory phenomena associated with a particle in movement, as well as its interactions with matter, represented by other elementary particles.

However, it is not absolutely necessary to enter into the particulars of these mathematical calculations to understand the fundamental aspects implied in the experiment of the two slots, which is considered one of the most appropriate experiences for analyzing the peculiarities of the quantum world. One of the key aspects of this experience is manifested by the fact that diffraction of particles that have two options of going through one of two slots only happens when we don't distinguish experimentally which one of the two possibilities occurs.

To define experimentally whether the particle goes through one slot or the other, we must modify a given experiment, by determining the position of the particle when it goes through the slot. For example, in the case of an electron, we can use a photon that collides with it and can be registered on a detector; in the case of diffraction of light, we can "mark" the photons by means of polarizing elements as they go through the slots.

This confirmation that determines the path of the particle implies modification of its wave function when collision of two particles (photon and electron) occurs or by the interaction with a polarizer. Thus, we are able to express the new interference function by means of the multiplication rule for successive events (the particle goes through slot (a) or slot (b) and a photon or a polarizer registers its passing). Empirically, we can express these interactions by means of coefficients, x_a ($0 \leq x_a \leq 1$) and x_b ($1 \geq x_b \geq 0$), that modify the amplitudes $\Psi(a)$ and $\Psi(b)$, giving place to a total function for interference:

$$\Psi_T = x_a \, \Psi(a) + x_b \, \Psi(b)$$

$$\text{If } x_a = 0 \quad \text{or} \quad x_b = 0; \quad \Psi_T = \Psi(b), \quad \text{or} \quad \Psi_T = \Psi(a),$$

and consequently there won't be diffraction. In experimental terms, we will say that when determining the path of the particle through one of the slots, we don't observe diffraction: The particle behaves as if going through one slot only.

E-3. "HOME DIFFRACTION" EXPERIMENT OF PHOTONS THROUGH TWO SLITS

It is necessary to wonder if this theoretical presentation can be carried out in the laboratory with such intellectual clarity. Recently, Rachel Hillmer and Paul Kwiat[1] published a simple diffraction experiment (done at home) of a photon laser beam divided into two fields by a very thin wire projected on a screen. The phenomenon of diffraction of the photons is observed very clearly on the screen by clear and dark bands (see Figure 26), when the photons have the option of going though either one of the two fields (left or right) defined by the wire, without evidencing which option was taken (case 1a). This situation is given when the two fields are free, defined only by the wire. Case 1b is given when we use a form for identifying each field in an equivalent and simultaneous fashion on both sides of the light beam. This is obtained by placing one polarizing sheet on each of the two sides of the light beam in parallel on each side of the wire.

If sheet 1 is placed to the left side of the wire, polarizing in a given direction, say vertical (v), and sheet 2 is placed to the right side of the wire, polarizing in a horizontal direction (h), diffraction does not occur.

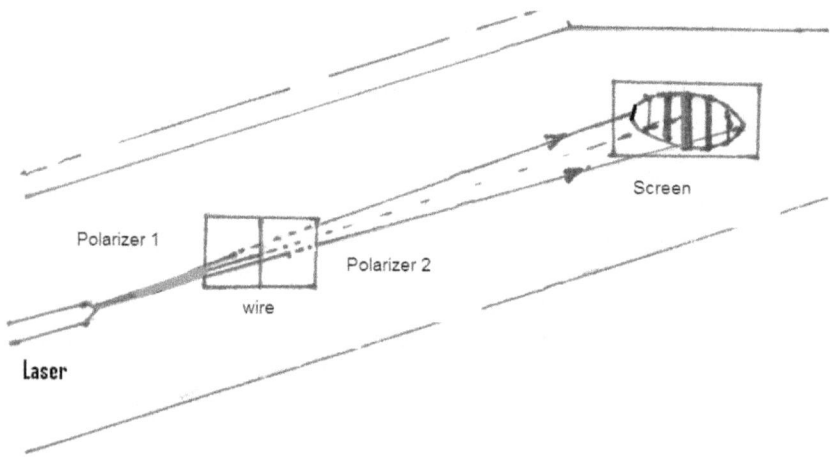

Figure 26: DIFFRACTION IN TWO FIELDS MARKING PHOTONS

Case 1a. Fields (left, right) separated only by a wire: Diffraction occurs. Case 1b. Polarizers 1 and 2, crossed fields (v and h): No diffraction occurs. Case 2a. Analyzing polarizer (v or h) placed between the wire and the screen: No diffraction occurs. Case 2b. Analyzing polarizer (v rotated 45°) placed between the wire and the screen: Diffraction occurs (Analyzing polarizers are not shown in this figure.)

---

[1] Rachel Hillmer and Paul Kwiat, "A Do-It-Yourself Quantum Eraser," *Scientific American* 296, no. 90 (May 2007): http://www.scientificamerican.com/article.cfm?id=a-do-it-yourself-quantum-eraser.

The way to mark the photons so they can be told apart is by means of polarizing sheets. For the reader who is unfamiliar with the phenomenon of the polarization of waves, particularly light waves, this refers classically to the position of the plane of vibration of the electric or magnetic vector (see Figure 19) that can occupy a defined position (for example vertical (v) or horizontal (h)) if the wave is polarized, or, to the contrary, the direction of the vector can vary randomly. The plastic transparent polarization sheet has the property of rotating the plane of polarization of light in a certain direction or determined axis. If we place two sheets successively with their axes crossed (perpendicular), they cancel the path of light through the two sheets, and individually each one polarizes the light in a determined direction (v, h), thereby allowing identification of a photon with a label (v, h).

Summarizing the cases presented up to now in the home diffraction experiment, we have:

1. Division of the photon beam into two fields by the thin wire:

   Case 1a. The wire is used solely: YES diffraction

   Case 1b. Employing parallel polarizers   YES diffraction
   1 and 2 (v,v) or (h,h):

   Case 1c. Employing crossed polarizers   NO diffraction
   1 and 2 (v,h):

   We interpret case 1a, which is equivalent to the classic experiment of two slots, this way: The significant fact refers to the dual choice for the photon that defines the experiment; thereby diffraction takes place.

   In case 1b, when marking the photon with the same label (v,v) or (h,h) on the two sides of the wire, we cannot distinguish its path by the left or right side; it is equivalent to case 1a.

   In case 1c, one can identify the path of the photon when passing through polarizer 1 (v) or polarizer 2 (h); therefore, there is no diffraction of the photon with itself.

It is necessary to wonder:

Does a classic interpretation exist for these observations?

Is it necessary to assume that each photon interferes with itself as it passes simultaneously through the two fields or slots?

Continuing with the home diffraction experiment, in its second stage:

2. We add to experiment 1c a third polarizer (an analyzer not shown in Figure 26), placed between the wire and the screen.

Case 2a. The analyzer has an orientation
v or h: NO diffraction

Case 2b. The analyzer v or h is rotated 45°
between v and h: YES diffraction

The third polarizer in case 2a causes the interference of the photon with itself to disappear after passing through the two slots. For this reason it is called the "quantum eraser."

The analyzer's effect in case 2b restores the diffraction phenomenon that had been eliminated by the polarizers v and h that had marked the photon when going by any one of the two fields (left or right). Again, in this case the analyzer acts as a "quantum eraser," making the two trajectories of the photon indistinguishable.

It is worth referring to the authors Hillmer and Kwiat in their own words: "This effect [cases 2a and 2b] involves one of the oddest features of quantum mechanics—the ability to take actions that change our basic interpretation of what happened in past events."

One of the peculiarities of quantum mechanics resides in the fact that a certain experimental behavior depends on what we try to find in that experiment.

In accordance with modern interpretation of quantum mechanics, interference happens when indistinguishable alternatives are combined in a given experience.

Generally, it is considered that the duality wave particle means that we can use one of the two models independently, as it suits us, to explain the phenomenon in consideration. Classically, we can interpret the diffraction of light through two slots exclusively as an undulatory phenomenon, just as we do with the diffraction of mechanical waves like sound, or with the waves on the surface of water. We can also use the electromagnetic model of Maxwell to explain the polarizing effect of v and h on the waves that go through the two slots with different polarization, which cancel each other out and therefore don't produce diffraction, as in case 1c.

Equally, we can explain the phenomenon of the "quantum eraser" as referring to the crossing of the polarization vectors of the waves from the classic model. In consequence, it seems unnecessary to use in this "home diffraction" experiment the quantum model of a particle that mysteriously interferes with itself. Does it follow, then, that all of these elaborate reflections are a deceit?

The home diffraction experiment that refers to a beam of photons is appropriate, by its simplicity, for showing the diffraction phenomenon under certain conditions compatible with both the classical and quantum theories. It has been necessary to carry out experiments on a single photon in interferometers of double beam[1] to conclude in a rigorous form, experimentally and theoretically, that the quantum model assumes the concept of wave-particle duality as a whole: The particle interacts with itself and, as such, it is described by means of the wave function that indicates an amplitude expressed in the complex plane of Argand (see Figure 25).

E-4. THE LIMIT BETWEEN THE QUANTUM AND CLASSIC BEHAVIOR; JOURNEY TOWARD A NEW CONCEPTUAL FOUNDATION OF THE QUANTUM THEORY

Complementarily, when diffraction is carried out with "massive" particles, such as the molecule $C_{60} F_{48}$ of a mass of 10^{-24} kg, equivalent to the mass of a million electrons, quantum rules have been verified using variable amplitudes of a wave function of the particle. This implies multiple superimposed or "entangled" states for each particle that interferes with itself. These experiences of diffraction of particles that come closer in size to our habitual world demonstrate to us the validity of the quantum model of material waves. However, when increasing the mass of the particles in movement, the interference lines start to vanish until disappearing completely.

It is interesting to reflect on the boundary between quantum and classic behavior. It is interpreted that quantum states that are superimposed cohesively are replaced by incoherent complex mixtures. If the two paths of the particle remain indistinguishable, the interferences will be visible; it is said that the states are entangled. On the contrary, if the steps are marked, that is to say, there is information on their state, the interference lines disappear.

Specifically, by increasing the mass of the particles subjected to diffraction, their quantum states of vibration and rotation increase their degrees of freedom; they become joined with their environment, and

---

[1]Paul Kwiat and Berthold-Georg Englert, "Quantum Erasing the Nature of Reality, or, Perhaps the Reality of Nature?," *Scientific American* (May 2007).

make the quantum effects of the system disappear as it passes into the classical world.

In the world of the philosophy of science, it is considered that quantum mechanics does not articulate fundamental principles, to the way of the axioms of relativistic mechanics, which are expressed by means of postulates like the principle of equivalence of physical laws in inertial systems, which are basic notions broadly accepted. In contrast, quantum mechanics uses axioms or mathematical rules as its founding principles, with their justification simply based on their utility in predicting experimental facts.

In the search for those general principles, which we denominated "metalogic" in Chapter 3, the philosopher Anton Zeilinger[1] suggests that the relationship between quantum states and elementary propositions, or "atomic" propositions, as Russell denominates in his principles of mathematical logic, corresponds to a "bit" of information, and consequently, one particular relationship can only give a defined state, in a specific measure. So then, molecular or compound propositions arise from the "entanglement" of the elementary states in a coherent form. For example, in the experiment involving diffraction through two slits, we will have "molecular" propositions that arise from considering elementary states of the particle cohesively with the two paths that the slits offer. These propositions define the quantum rules to express wave functions, in the form to which we have referred.

It is necessary to add that the quantum world defined in the Schrödinger equation is deterministic; the wave function defines all its values for x and t, and consequently the values of the amplitudes of the wave are defined states. However, these elementary states lose their coherence in complex systems in such a way that indetermination is introduced. Similarity of these phenomena of quantum mechanics, with the interphase between deterministic and chaotic states described in nonlinear dynamics, opens a promissory future in the frontier of science.

5-6-3 The End of Science or a New Paradigm?

Chemistry arose from two conceptions or hypotheses about the diversity of matter. Fortunately, these concepts subsisted for two thousand years, when a new emergent order gave place to the transformation that we call scientific. It contained implicitly Karl Popper's ideas that a theory can-

---

[1]Anton Zeilinger, "A Foundational Principle for Quantum Mechanics," *Foundations of Physics* 29, no. 4 (1999).

not be a proven reality, but rather, a believable fact by means of its confrontation with experimental facts; and yet, as numerous as these facts may be, while they can increase a theory's credibility, they don't constitute in themselves an absolute test of its truthfulness. All theories called scientific, because of their nature, are "falsifiable"; that is to say, not being complete, their limitation implies they can be substituted totally, or introduced in a contextual order of higher hierarchy. Those hypotheses that don't meet this limitation of "falsifiability" are in themselves irrefutable and are outside of the realm that we call science. Bertrand Russell proposes an illuminating hypothesis of this non-falsifiable type: "The world was created five minutes ago." There is no way of proving the opposite; there will always be an explanation that discards any argument against it, without being able to prove this statement wrong or falsifiable. We can argue, for example, that the world was made this way, just as it is, so that we would believe that a past exists previous to creation—which happened five minutes ago—and this is simply what it looks like.

These ideas show us clearly how science can become pseudo-science. Are natural sciences reaching a point where they are becoming non-falsifiable? Frequently ideas regarding the end of the physical sciences are published,[1] highlighting the impossibility of advancing within the current theory that has locked itself in. The cosmology of the big bang theory (see Section 4-2) has inklings of returning to the Christian theology of creation. Let us remember Einstein's reaction to Father Lemaître when he proposed for the first time the "primitive atom" as the origin of the universe.

Authors, such as Stephen Hawking, who are adherents of the current scientific paradigm, tell us that time began fifteen billion years ago; it doesn't make sense to question if there is a *before* before the instant of creation. That is outside of the realm of science. This may lead us to believe the current theory is no longer a scientific theory, because it is not falsifiable. If multiple universes are proposed, considered in extra dimensions, these would be outside of our observation, because the same theory proposes a closed universe. Could it be that our scientific theories that have their origin in the search for the "laws of nature" are returning to their mythological base, the Judeo-Christian paradigm?

In the same way, modern atomic theory, which originated from the hypothesis of the Greek "atomists," arrives currently at a conception of matter that unavoidably returns to the simplicity of its origin. It is said

---

[1]David Lindley, *The End of Physics* (New York: Basic Books, 1993).

that science cannot go on further if we delve into speculative territory. Let's reflect on this: Science is essentially speculative; its theories should be falsifiable. Ideas like those presented by David Bohm and F. David Peat in their book *Science, Order, and Creativity,* and by Ilya Prigogine, to whom we have referred, stimulate changes to the current paradigm that seems to be drained. Concerning the term *paradigm,* it is worth observing that it doesn't designate the same idea expressed by firmly defined theories in current science; it refers to a "subliminal" framework of knowledge that forms part of the collective unconscious that as such is not expressed symbolically.

The Aristotelian conception that we have analyzed in connection with the epistemology of chemistry was part of that conceptual framework, not expressed symbolically, that characterized the paradigm prevalent in the time of medieval Europe. Its evolution doesn't seem to have ended; on the contrary, it is arriving at a crucial point that is expressing itself as a new synthesis called systemic, in opposition to the reductionist paradigm that has prevailed up to now.

Few references are made to the importance of the history of chemistry in the evolution of the current scientific paradigm; the victory of atomism is a very significant part of the popular ideology of our time. The concepts called materialistic that characterize the contemporary cultural paradigm are much closer to the original conception of chemistry than to the mathematical-physical theories, for obvious reasons of their conceptual simplicity that has come to be part of the subliminal framework of knowledge. This may be one of the reasons very little is written today about the foundation of chemistry; it has become a part of the collective unconscious that is a given and is incontrovertible.

Physical-chemical models have impacted biology, which evolves experimentally and theoretically with an extraordinary dynamism that reminds us of the same path upon which chemistry once traveled. Science studies progressively experimental facts of greater complexity, growing rich initially in a qualitative way, as it opens the way to advances in its physical-mathematical interpretation, which evolves from analytic observation toward the systemic understanding that contributes more rigorously to the study of the complexity of nature. Two philosophical conceptions of science are present in the new century. In good part, the scientific traditionalists prefer to ignore refreshing breezes that seem to be turning into storms.

The scientific branches close to empiricism, such as chemistry and especially biology, which at the moment is beginning a revision of its foundations, seek an interpretation of natural complexity that points to indetermination, in opposition to the causal linearity of traditional science. Cautiously, renowned scientists give little importance to the impact of the new mathematics, designated as the theory of chaos, or the analysis of nonlinear dynamic systems, reducing their influence to classical mechanics and omitting any analysis in connection with quantum mechanics that scientists such as Bohm and Prigogine, among others, have firmly proposed.

The phenomenon of life represents a continuity in universal evolution; its integration with physical chemistry began with the development of carbon chemistry, also called organic, which extended its methods for synthesis to products that were thought to be exclusive to the biological world. Biology today redefines its complexity by means of theories that extend statistical thermodynamics, chemical kinetics, and other disciplines to systems of a larger complexity that demand novel mathematical treatments characterized by their "circularity" and designated mathematically as nonlinear.

It is about the interpretation of facts within complex systems that are neither classical nor quantum, words that refer to physical theories and not to facts.

On the other hand, it becomes increasingly evident in the contemporary cultural panorama that what is needed is a strong interdisciplinary group that proposes new directions to the development of human knowledge, by means of a new synthesis that interprets Western science and perennial philosophy in a wide context, and that is integrative and constructive.

5-7 BIOLOGY: THE FRONTIER OF SCIENCE

Just as chemistry, starting with Aristotle's Greek philosophy, revived the ideas of the old atomists by means of the new scientific paradigm of objectivity, scientific biology abandoned the traditional concept of the duality of spirit and matter of the scholastic metaphysics, a synthesis of Plato's philosophy and of the Hebrew mythology. The new science, when studying live matter, followed an analytic method similar to the one adopted in chemistry, assuming a universal continuity expressed in the theory of the evolution of species starting from matter called inorganic or mineral.

Darwin's theory of evolution gave a historical dimension to the new science that had remained basically in a restricted environment of descriptive classification of the phenomenon of living matter. By means of this theory, biological diversity registered in the new geographical explorations acquired a new dimension that looks toward the past, suggesting a material unity of the universe.

5-7-1 The Scale of Time and Biology

The universe, our world, is not stationary or homogenous; changes exist and there is diversity. It is in this diversity that the origin of knowledge resides, which relates phenomena and events by means of communication between systems, such as our brain and the external world. That diversity is expressed in science by means of the inseparable concepts of space and time, as has been demonstrated by the theory of relativity. Likewise, our individuality, our beloved "me" or "ego," is a by-product that emerges from that communication that is life itself. Strictly speaking, life exists if there is diversity and communication, that is to say, exchange of information. From that point of view, life in our present reflects the universe's whole existence. This present time that we call "real" is also local; it is bound to space, and both are related indissolubly. What we call past emerges from comparing communication patterns between regions of our brain that establish the memory and originate the "me" concept.

In this order of ideas, we refer the reader to Chapter 3 on symbolism and order, specifically to Section 3-4, where some ideas are presented on the establishment of the scale of time in our mind. By way of summary, we can state that the order of registered events in our memory and, on a smaller scale, in animals, is a reflex of that sequence in which some phenomena occur in nature, repeating with the same ordering pattern. These sequences in nature give us the initial rule that allows us to create the scale of time that is expanded by means of the symbolism of language.

It is worth contemplating whether these stages that establish the time dimension in our mind are also materialized biologically. Does some biological clock that has been evolving from our most remote ancestors exist? We could speculate that the biochemical reactions that happen in our cells following defined sequences are probably a genetic base for the neural establishment of the scale of time and its arrow, in a way that we designate as instinctive. Neurological studies[1] show that those

---

[1] Rodolfo R. Llinás, *I of the Vortex: From Neurons to Self* (Cambridge, MA: MIT Press, 2002).

biochemical sequences that occur at a scale of fractions of a second are a fundamental part of the electric stimuli that coordinate the muscular contractions responsible for organized corporal movements. It is considered that control of organized movement led in animals, in contrast with plants, to the generation and nature of the brain whose main function is prediction; in other words, the expectation of sequentially ordered experiences. In this way the concept of time expands to what we call the future.

The existence of the ego implies isolation of a region of space that sustains itself, maintaining communication with the rest of the universe. Strictly speaking, that "me" is the relationship between those two regions, one cannot conceive of but in holistic form—that is to say, both as a group. From internal communication emerges the past, which assumes registering in one's mind a thought that when established in repetitive form constitutes a habit, leading to a projected expectation that we call the future. In contrast, in the universe in general, only the present exists; time emerges with life, without which it is inconceivable. In what we call inanimate nature, there is diversity, communication, and change, but there are no regions that sustain themselves and that, with their continuity, permanency, and differentiation, give place to knowledge that is none other than the external and internal communication of the phenomenon we call life.

The evolution of the universe in all its transformations culminates in life, in our present, which is the only existent thing and which is also past and future; simply because it is this way, it is a fact. Expansion of the scale of physical time to cosmic dimensions is a product of evolution at a social scale; it is proper to mankind that extrapolating the knowledge we call scientific projects us to the confines of space-time. When we observe light through sophisticated telescopes, we can distinguish regions of space located at distances never imagined; scientists, by means of elaborate theories, interpret those electromagnetic signals as coming from the first stages of the evolution of the universe, assigning them to their remotest past. It is worth asking ourselves what validity we can assign to these explanations, inferred from data that reaches our brain by such elaborate means.

The answer to this question is relatively simple; all that we call reality, from the existence of matter at a macroscopic level to subatomic particles or the distant quasars, is nothing more than an explanation, smaller or larger, mentally generated from data that our brain receives through our senses. To ask ourselves if this is real or not is to pose a question that cannot be answered by science; it is outside of its realm. Only the philosophical approach to the meaning of knowledge can guide us to elucidate its validity. If our attitude is positivist, we will say that the complex mathematical formulations elaborated by modern science are consistent with experimental data: That is to say, its predictions have been confirmed experimentally, independent of which may be the models or interpretations that may be given to the theory in the realm of our habitual world. If we are realistic, not necessarily friends with the king, we will say that an external physical universe exists that we explain by means of a model that comes closer to our habitual world, allowing us to live within the parameters of practical reason, and that in turn is consistent with elaborate scientific theories that are subject to revision and evolution by principle. If we are solipsists, we will say that only our mind exists and that this alone is reality. If we are modern creationists, we will propose that a supreme being created the universe just as it is, without denying its evolution, and hence we come closer to the realistic model as long as it doesn't oppose the particulars of our faith. If we are idealistic, we will say that the only real things are the universal ideas present in mathematics, logic, and aesthetics.

Looking at another option, we come closer to the perennial philosophy of mysticism by recognizing that without the concept of biological time and its origin, there is no knowledge because there is no diversity, no individuality, and no existence. In our life, there are dimensions that are not framed in space-time. We feel them; they are not part of reason, they cannot be expressed by means of spoken or written symbols, they transcend our ordinary life and are intimately bound with our happiness, which manifests itself when we forget about the "me" and about time that makes our existence possible but also is our yoke and our suffering. Is it possible to live without time? Without suffering? Buddhism and in general Hinduism respond affirmatively, advising us to suppress our desires for the future, forget our past, and be reborn—to be children again and transcend our life while ascending in the scale of true knowledge.

5-7-2 Toward a Theoretical Biology

The search for a theoretical base for biology is a novel scientific area of great importance to the consolidation of knowledge from a holistic point of view. Converging in this task are philosophical reflections, mathematical-physical methods, and, in a more specific way, recent experimental advances in biology and chemistry.

Actually, biology is the frontier of natural science; it is evolving quickly from the experimental point of view with its extraordinary discoveries that impact medicine and all the realms of human life. From the theoretical point of view, biology looks for its foundation in physics and chemistry and in turn generates new concepts or theories that, with their "circularity," revert to traditional science.

The Web of Life, an excellent book by physicist Fritjof Capra,[1] divulges ideas that have as a mathematical base the theory of chaos, which refers to the study of singularities or bifurcations of complex systems that intervene in network effects, which are considered cause and effect mutually at each one of their singularities or bifurcating nodes. These mathematical models that are denominated "nonlinear" have found applications in such fields as meteorology, economics, and biology. Theoretical biologists, such as Chileans Humberto Maturana and Francisco Varela,[2, 3] erupted onto the international scientific scene, distinguishing themselves with their novel contributions to science and philosophy, by means of a new concept for the characteristics of the phenomenon of life, called "autopoiesis." This concept expresses the autonomy and self-reference of the vital units, whether in the microorganisms that sustain themselves in their individuality or in the multi-cellular organisms like man, who characterizes himself by his hierarchic self-organization.

One of the most relevant aspects of the mathematical theory of complexity or nonlinear systems implies a new concept for indetermination that radically differs from randomness; it doesn't refer to the occurrence of "punctual random" phenomena, such as those of the game of roulette, but rather to the bifurcations of the states of complex systems that are characterized by remaining inside dynamic frameworks that succeed each other in their characteristic trajectories or basins.

Individual phenomena that occur randomly, such as natural decomposition of radioactive atoms, follow a mathematical law that allows, for example, for statistical calculation of the probability of natural decomposition of a radioactive sample that implies a high number of atoms. In contrast, natural phenomena, such as those that refer to meteorology, involve complex interactions in which a very large number of variables characteristic of the system intervene. They are expressed mathematically by means of nonlinear equations whose numeric solutions cannot

[1]Fritjof Capra, *The Web of Life: A New Scientific Understanding of Living Systems* (New York: Anchor Books, 1996).
[2]Humberto Maturana, *La realidad: ¿objetiva o construida?* (Anthropos Editorial, 1997).
[3]Francisco Varela, *El fenómeno de la vida* (Dolmen Ediciones S. A., 2002).

be calculated by means of simple algebraic equations, but by repetitive methods of numerical adjustments that demand the employment of modern computers, which transport us with their high speeds to times comparable with our short existence.

From a philosophical point of view, in the search for theoretical principles of biology, it is important to review the meanings of some terms that have been considered relevant in answering such questions as: What is life? What are the characteristics of living beings that differentiate them from the rest of the universe? Equally important, once we've established general semantic bases, is the deeper search for scientific principles that are coherent with experimental data and that can be projected toward new frontiers of research.

5-7-3 Some Relevant Considerations in Biology

A. NATURAL AND ARTIFICIAL

The analysis of these two words begins the semantic and philosophical considerations of Jacques Monod, in his writings on the natural philosophy of modern biology. Traditionally, the sciences called "natural" have evolved toward physical sciences, understanding that mathematical-physical principles extend progressively from simple systems toward the complexity of chemistry and biology. The name "basic sciences" has also been coined to differentiate them from "human sciences," a term referring to the interrelations in human society. As in all divisions, their limits are not clear: Where do we place psychology or medicine?

The word *natural* incites certain reverence, certain magic; what is natural is distinguished as coming from a spontaneous process, whereas an artificial thing supposes a preestablished action. In physical chemistry, the spontaneous is established by thermodynamics, in which an analysis is made by means of considerations of actions between systems, regions selected arbitrarily that define their complexity, or multiplicity considered, which allows evaluation of some numerical variables, such as entropy, indicating the grade of order or disorder of such a system. In this scientific context, the word *natural* is synonymous with spontaneous processes associated with increments of entropy that in turn correspond to evolution. The second principle of thermodynamics establishes the increase of entropy in closed systems, such as the universe, postulating its natural evolution from order to chaos, and suggesting an inherent contradiction with the beginning of life that we interpret from a molecular point of view in an inverse direction. However, this contradiction is apparent; thermodynamics, when defining one system, simultaneously defines another one that corresponds to its surroundings. Consequently, the entropy increase refers to the net increase for both systems: Where life is created, entropy is reduced, and in its surroundings it increases in a higher proportion, resulting in a positive net increase of entropy in the universe.

In terms of our habitual language, when arbitrarily we don't include ourselves in "what is natural," we are external agents that produce "what is artificial." Otherwise, we would have to assume an external cause to the system of nature, with the purpose of explaining the artificial. The acceptance of this semantic duality leads to an attitude that favors the hypothesis of God the Creator, who, by means of His intervention, the laws of nature are established. On the other hand, the historical development of science shows us that the election of natural systems—that is to say, those that are not defined arbitrarily, but on the contrary are defined by means of observation—leads us to coherent hypotheses with experimental data. In physical chemistry, the search for the atom of the Greeks produced molecular systems that break down into other systems such as atoms and subatomic particles. In a similar way in biology, the discovery of cells and bacteria established natural systems that we should choose as basic units in the causal relationships. This analytic methodology, the basis of modern science, has been denominated in a pejorative way as "reductionism," because of its opposition to an attitude that involves the totality, which is frequently designated as "holistic."

It is evident that the election of a system to be studied is indispensable in the analytic method, as its name indicates. The criticism of reductionism is based on its use of causal relationships, called linear relationships, that contrast with feedback effects, or, in other words, with circularity. This latter systemic consideration served as the basis for the holistic conception, which considers the system and its environment jointly in all its aspects. With the purpose of establishing these considerations in mathematical form, nonlinear algorithms have been proposed that have had an extraordinary development in recent decades, reminding us of the works of Henri Poincaré at the beginning of the twentieth century. Their applications in biology are of the greatest importance, as they are in disciplines that at first sight seem very distant, such as economics.

B. INDIVIDUALITY AND IDENTITY

With an eye toward a higher semantic rigor, it is convenient to define these colloquial terms clearly that naively can be considered synonyms. Individuality refers to the establishment of a system or region that is generally considered bounded by spatial dimensions. However, it is convenient to generalize the space concept to other variables of all order, thereby introducing the so-called "space of phases" or multidimensional space (see Figure 27). This mathematical artifice allows the generalization of the concept of intuitive space in three dimensions, our home reference, to a space that is not in the imagination, but that is of great conceptual utility for characterizing systems globally in their evolution. In this way one can choose a system of (n) atoms, molecules, or microorganisms that have a number (v) of certain characteristics, indicating their state by means of a "point" in the multidimensional space with phases of (v x n) variables, and visualizing their parametric variation with relationship to time by means of a line.

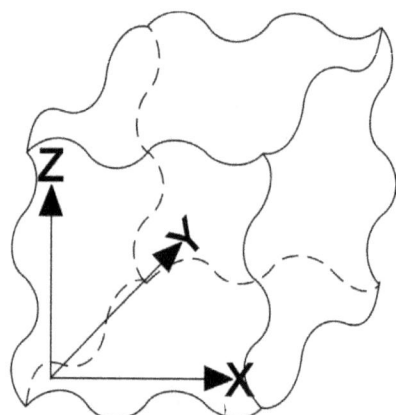

Figure 27: THE N-DIMENSIONAL SPACE

Identity scientifically expresses the equivalence of all variables that define the systems considered, excluding the space variables, because it is assumed that identical systems occupy different places. In daily language, identity usually refers to characteristics of objects; in biology, we consider live beings, from one-cell microorganisms, to the immense variety of multi-cellular beings such as vegetables, to animals that we call superior, certainly including ourselves within the latter category.

Frequently, the term *identity* is restricted to humans, because implicitly we find that it doesn't have relevance for other beings. Obviously, we can state that twins acquire their individuality in the cellular division that originated them. We assign to these twins the same identity because their genetic characteristics are the same, nevertheless recognizing that circumstantially they acquire different identities as they evolve with age. These semantic differences are important when we want to be rigorous and place ourselves on firm land when dealing with theoretical bases. Those subtle differences between "individuality" and "identity" are relevant in connection with the importance that "vital units" have in originating the phenomenon of life, which cannot be feasible without the spatial differentiation of those fundamental units whose basic characteristics are its autonomy and its sustainable evolution.

Those vital units are possible physically by means of the membranes that individualize a region, giving these units their vital characteristics, and allowing simultaneously their isolation and communication with their environment. Isolation provides them with their individuality that gives them their temporary permanency and the possibility of evolving toward other identities as they interrelate with the environment. In consequence, individuality—that is to say, the constitution of units differentiated spatially—is a fundamental characteristic that allows for life, without which it is inconceivable; however, it cannot be considered isolated from the "prebiotic soup" that made possible its formation, with which it constitutes an indissoluble vital system.

Dynamics of this system formed by the vital unit and its environment are characterized by two trajectories, metabolism and evolution, that exist in very distant time scales. Metabolism, which constitutes the system's vital scheme, makes possible its immediate self-sustainment, and evolution projects its survival to conquer time, achieving "immortality." In both processes an interrelationship exists between the being and its environment at very different time scales: Evolution can be designated as the gradual transformation of the identity and metabolism to the self-sustainment of its individuality.

In 1943, similar considerations led Schrödinger,[1] in his famous essay, "What Is Life?," to propose theoretically for the first time the genetic units as stable macromolecules that replicate and transmit the fundamental characteristics responsible for the identity of cells, evolving by means of changes or mutations that occur in a quantum fashion, overcoming high-energy barriers that lessen their probability. In contrast, the mechanisms of biochemical reactions that characterize metabolism use

---

[1] Erwin Schrödinger, *What Is Life? Mind and Matter* (Cambridge: Cambridge University Press, 1988).

catalytic enzymes that facilitate their vital cycles, which occur at a much smaller time scale than do genetic changes. Undoubtedly, this theoretical essay contributed notably to the discovery of DNA, and set a new direction toward the establishment of a theoretical biology.

C. THE MYTH OF THE "EGO"

The concept of "me" traditionally refers to its psychological meaning, to a unified sensation that is completely private and global, in contrast with the sensations that come from our environment. Physics, in its general form, is derived directly from the information obtained from our senses, constituting private or individual facts that, by consensus with our fellow humans, define facts with universal or public value, expressed symbolically by means of language.

Most scientists, especially biologists, adopt a materialistic conception of the "me," based only on physical data. Specifically, neurologists consider that the relationships between the physical observations of the brain and human behavior reduce the concept of the "me" to the physical-chemical phenomena within the brain. For Rodolfo R. Llinás,[1] the myth of the "ego" collapses with the knowledge of the neurological and biochemical mechanisms at a cellular level, and by means of the electromagnetic images that show cerebral regions that seem to intervene more directly in a certain psychic state. The globalization and coherence of the activity of neurons are considered to manifest themselves physically by electric oscillations of 40 Hz generated by the brain. It has been suggested that this characteristic activity of the nervous system is responsible for the sensation of the "me," or consciousness. The physicist Roger Penrose[2] suggests a quantum entanglement as a possible phenomenon of cerebral coherence that is implicated in consciousness.

Whatever physical mechanism may be implied or associated with the subjective concept of the "me," and without diminishing the scientific importance of all those studies, it is not reasonable to reduce the entire mental phenomenon to the related or coexistent physical experiences of the mind. The arrogant attitude of many scientists contrasts with that of philosophers like the Dalai Lama,[3] who, without pretending to be a deep scientific thinker, considers that reality is not satisfied by or reduced to physical knowledge. Buddhism, consistent with many Western philosophers such as

---

[1] Llinás, *I of the Vortex.*
[2] Penrose, *Shadows of the Mind* (see Chapter 3, Section 3-2).
[3] His Holiness the Dalai Lama, *The Universe in a Single Atom: The Convergence of Science and Spirituality* (New York: Morgan Road Books, 2005).

Bertrand Russell, as well as the context presented in this book, proposes three fundamental categories of knowledge, according to their source:

1. Physical knowledge that is public and that we call material or objective
2. Knowledge based on subjective, private experiences, such as psychology and mysticism
3. Induced mental elements that we call abstract, such as mathematical logic

Platonism considers the theoretical mental elements as having an independent self-existence, and as a pre-conditional basis of the material universe—a position that is shared, at least sentimentally, by mathematicians such as Penrose who admire the sublime beauty of the superb principles of physics and mathematics. In this respect, the position of this book approaches Buddhist ideas, which consider that the mental environment cannot be reduced to the world of matter, although it depends on it to be able to function, as well as those of Platonism, which values mental abstractions as immortal, though moderated by the recognition that these come from the relationship between mind and environment.

An area of biology and neuroscience currently being developed with broad-ranging perspectives refers to evolution of the nervous system in the context of embryology and the evolution of the species, including our most distant ancestors, the bacteria, which evolved in the primitive oceans during thousands of millions of years.

The expression the "myth of the ego" that we want to analyze in this section doesn't refer to the physical conception outlined by neurology, but to its psychological aspects and its existential meaning. However, it is necessary to wonder: What is the meaning of this oppressive word, *ego*, in connection with biology? Or, on the contrary, can its physical meaning be expressed more appropriately by the word *individuality*?

Biology, as a natural science, is based on the so-called principle of objectivity that by its own denomination is not provable; it constitutes a methodology that is considered indispensable in the process of the acquisition of knowledge. In the physical sciences, considered in their more general form, objectivity refers to the observable facts, those phenomena that affect our senses directly or by means of instruments. From this point of view, in biology, the so-called "animism" that assumes a force or vital principle as the cause of life in its origin and development, negates itself by principle. The justification of this scientific methodology is strictly based on its utility in the search for knowledge, in contrast with mythology that may have a social, psychological, moral, political, or any other kind of value, but is not appropriate for acquiring knowledge about nature—that is to say, chemical, physical, or biological.

We may ask: What is the validity of the principle of objectivity, fundamental to the natural sciences, as it is extrapolated to other areas of knowledge, such as psychology? The social success of the natural sciences, and their general acceptance, is owed especially to their technological consequences, and not properly to the scientific knowledge in itself that largely remains ignored by most of the population. Despite this fact, paradoxically, the scientific method and its principles have been generalized to areas very distant from physical objectivity such as human sciences. Strictly, the methodology of natural sciences cannot be translated entirely to all areas of knowledge, where the observable events cannot be established in a clear and objective fashion; on the other hand, the employment of mathematical symbols, which are fundamental in the logical development of physics, lose their value at least partially in other areas of knowledge.

Evaluation of knowledge also depends on the philosophical attitude that is adopted in its interpretation, which is independent of the scientific method. When in biology we seek to extrapolate biochemical or neurological knowledge, and relate it to psychology or mysticism, we should be cautious and remember that psychological events are by nature different from physical, observable ones. If we affirm that we have understood the mystic experience or the value of deep meditation, because we have detected by means of sophisticated instruments certain registrations or patterns that happen simultaneously with these psychic states, we are not contributing anything new to the area of psychology.

In Chapters 6 and 7 of this book, philosophical, psychological, and mystic aspects regarding the nature and transcendence of our consciousness are proposed. In the environment of biology, we emphasize the physical "me" that emerges as a basic element of life that remains from its origin in the first unicellular beings that evolved within themselves during enormous geologic periods, consenting recently to the pluricellular beings, well-known as plants or animals. Our beloved "me" constitutes the maximum self-centeredness that nature has imposed to preserve the continuity of life on planet Earth. Physical individuality has evolved toward human consciousness, our yoke that ties us to physical nature, while at the same time allowing us by means of spiritual transcendence to liberate ourselves from our material dependence.

In this sense liberation from the myth of the ego is begun by recognizing its origin and meaning in the initiation of life and its continuity. Individuality, the "me," is imprinted in our nature; it is the price we have to pay for our existence, and as such it is a means that we should use wisely without allowing it to enslave us. Let's contemplate that the Darwinian competition for life has not been the only evolutionary force implied in natural selection; bacteria were ahead of modern genetic engineering by thousands of millions of years, exchanging bits of genetic information by means of fragments of DNA at a speed that approaches modern communications. In this sense, Lynn Margulis and Dorion Sagan state that in the "biological microcosms the implications of rapid genetic exchange result in that one cannot strictly speak of true species in the bacterial world."[1] All bacteria are organisms, entities able to use genetic engineering on a planetary scale.

Understanding these biological facts bears the modification of our current paradigm, which considers the evolution of life like the law of the jungle or the survival of the fittest; this paradigm originates in a reductionist analysis that contrasts with the complex processes that have made possible the evolution and sustenance of the biosphere as a whole.

The Judeo-Christian mythology on the creation of life, which today is designated as creationist, considers the "needed being" not only acting in the initiation of life, but also individualizing His intervention in the creation of each species. The new science that challenged, by means of objective observation, the prevailing religious dogmas reached its initial heights with Newtonian and Galilean mechanics that debunked the egocentric conception that located man and his home in the center of the universe. In an area of knowledge with sensitive moral implications, the studies and observations of life found greater difficulties in conquering the established dogmas. European naturalists of the seventeenth and eighteenth centuries began their observations of fossils, interpreting their origin timidly in very near form to the biblical context. However, expansion of the geographical and naturalistic explorations propitiated the emergence of the evolutionists who proposed, sometimes in anonymous form, that God had created life a single time, without intervening later in each one of the animal or plant species that had evolved spontaneously. Charles Darwin, nevertheless, who in the beginning was attached to the biblical stories, at least publicly, gathered a group of biological observations that led him to express to some of his colleagues sentences like the following:

---

[1]Lynn Margulis and Dorion Sagan, *Microcosmos: Four Billion Years of Microbial Evolution* (Berkeley, CA: University of California Press, 1997).

Lastly, luminous gleams have arisen, I am almost convinced, contrary to my previous convictions that species are not inalterable.

After overcoming significant moral and psychological difficulties, even within his own family, in 1859 Darwin published his famous masterpiece, *The Origin of Species*, which proposed the theory of natural selection. Even though evolutionists like Jean-Baptiste Lamarck had previously proposed theories on the evolution of life, these were not substantiated fully by established experimental facts like in Darwin's work. On the other hand, Lamarck's theory refers to a "vital force" present in biological evolution from its initiation that directs the adaptation of life to the changing environmental conditions. In contrast, in Darwin's theory the mechanism for the evolution of the species is explained by variations of vital singular characteristics that occur in a fortuitous way and that are selected collectively in the species, by means of the survival of the most capable individuals.

The validity of Darwin's contribution to the theory of biological evolution is based faithfully on the observation of nature, in contrast to its predecessors, which lacked objectivity. On the other hand, the dynamics of Darwin's evolution of the species, based on the "survival of the fittest" or "natural selection," is in itself a statement that is designated logically as a tautology. Let us remember the logical-philosophical writing of Wittgenstein, mentioned in Chapter 4, which can be summarized as: Tautology and contradiction are not figures of reality but belong to symbolism, as zero is in arithmetic. Tautology is unconditionally true and contradiction is not true in all conditions.

Let us pause; specifically, "survival" and "fittest" mean the same thing. Then, what value does this statement or proposition have? Certainly it is not a false statement; it simply brings to light an equivalence that is useful for logical development. Natural selection states a mechanism for evolution, only if we define how the biological singular transformations occur. In Darwin's time biology referred to the "macrocosmos," to the living beings such as animals or plants, with observable characteristics to the scale that our senses would gather directly. The hypothesis proposed by Darwin, which considers that the variations of the characteristics of the species originate individually in a fortuitous way, could not be verified experimentally in its time; seemingly it appeared contradictory to the immediate reality. During one's lifetime it is difficult to observe transformations of biological characteristics that characterize species.

The success of this hypothesis had to wait for the instrumental advances that allowed for observations of the biological "microcosmos" of bacteria and cells, as well as the development of genetics at a molecular level.

The discovery of the structure of the macromolecule called DNA, one century after Darwin attempted to establish a causal relationship between the sequences of nitrogenated bases, found in this "molecule of the life" the characteristics that make live beings differ, whether in the environment of microorganisms or in pluricellular beings like man. On the other hand, it has been considered in modern times that there is a low probability that mechanisms that modify macromolecules of DNA and RNA will occur in nature in a fortuitous way, due to high-energy barriers that are involved in these transformations. In this context, the neo-Darwinian theory is presented; it establishes that these molecular transformations produce changes or genetic mutations that don't lead to qualities that in a large part are favorable to adaptation of live beings to the environment. However, a fraction of these mutations will give place to more gifted individuals, who will be selected in their struggle for existence, leading to the evolution of the species. Particularly in the microcosmos, these stages and mechanisms of evolution can be verified in an accessible time scale to our individual lives.

Extrapolation of biological theory in studying human social groups, unfortunately, has "justified" the formulation of rigid ideologies such as economic liberalism, which in its extreme form gave place to the excesses seen in the early years of the Industrial Revolution. These, in turn, led to Marxist ideology and, later in the twentieth century, to communism that idealizes a society without singular differences, without economic classes, governed by "angels" who are assumed not to be subjected to the biological competition inherent in human nature.

Dynamic and flexible cooperation is the key that preserves and modifies the biosphere. In the microcosms that evolved during two billion years in the first stages of life, the genetic exchange is implied among bacteria as a fundamental mechanism of evolution that coexists with genetic mutation. All bacteria behaved as a single organism, like an entity that was ahead of the emerging genetic engineering on a planetary scale. A bacterium never works as an isolated individual; groups of different classes of bacteria live collectively, responding to and reforming their environment.

The "law of the jungle" or "natural selection" is professedly reductionist. Immortality is the web of life as a whole. Individuality and its ultimate expression, the "me," traps us in time, between life and death.

Liberation begins by understanding that the myth of the "me" is only a means that is imposed on us by nature for our existence. The continuity of life has been achieved by means of a balance between the competition that preserves the individual and the cooperation in ecological niches that stabilizes the entirety of the biosphere.

D. CHANCE AND NECESSITY

These words, which Jacques Monod[1] used to title his cited book, signify a polarization of the attitudes that have characterized the interpretation of the history of biology. Their meaning has been revised in terms of the indetermination of systems far from equilibrium that differentiate the fortuitous, or random, events from bifurcations among basins defined by patterns that characterize the "chaotic" systems.

Random or chance assumes all the options possible for the player from the whole deck of cards, unlike the indetermination represented by non-linear algorithms that are used in general to simulate properties of systems characterized by their "circularity," such as biological systems. These in-determinations are limited to some bifurcations of the dynamic state that we can compare to the card player's situation, when in a game with others the number of cards available decreases in contrast to when he plays alone. The game of life is not that of a solitary player in a fixed and possibly hostile environment; an interrelation exists among each player and the environment constituted by the distributed deck of cards among all the players. Which resembles life more: playing roulette or a group game such as bridge or *tresillo*? (*Tresillo* is a Spanish card game with three players where two ally themselves against the last single winner of the pot.)

The property called circularity establishes in biology a fundamental element to bring us closer to the complexity of the phenomenon of life. If we restrict ourselves to the duality of the individual system—that is, one's self and, on the other hand, the environment or our surroundings—circularity refers to the reciprocal influence of these two entities that are affected mutually. It is worth remembering the myth of the Eastern religious philosophy that is symbolized by the snake that devours itself, gobbling its own tale, or the symbols of yin and yang, which represent the interrelation of cause and effect between the being and his environment. On the other hand, the environment can be constituted by the universal physical medium or refer specifically to vital systems that we consider internally as part of the being according to our focus.

---

[1]Jacques Monod, *Chance and Necessity* (New York: Vintage Books, 1972).

These considerations are of great importance in connection with the evolution of life and its origin. Lamarck, prior to Darwin, visualized evolution of the species by means of an unspecified mechanism that highlighted as a creative force the direct influence of the environment on living beings. In this context, the duality of chance and necessity is manifested in subtle form in both theories. Concretely, in these theories the action of the environment differs, either in a causal, direct form or indirectly by means of natural selection.

New scientific directions have arisen to which we have referred as complementary to the so-called reductionism that are designated as "systemic"; these imply a causal circularity that opens new horizons in scientific and philosophical investigation. In particular, the post-Darwinian theory of evolution has a more flexible and less dogmatic philosophical position than that adopted by the so-called materialism, the product of an extremist reaction against the Western traditional religion that was opposed to modern science.

The ideological extremism that simplifies the environment of thought brings us close to electronic computers that only distinguish the duality of yes and no, by means of the numeric binary symbols 0 and 1; in politics they classify us as "right-wing" and "left wing," in the philosophical-religious arena we don't have another option than to be spiritualistic or materialistic, and in biology we must choose between chance and necessity. The phenomenon of life is neither necessary nor a product of chance.

The laws of physics, starting with Newton, formed deterministic thinkers that exercised a great influence in all the arenas of knowledge that were popularized in the twentieth century, giving to science the character of a new dogma or paradigm. The collapse of the Western religious ideology secured the so-called materialism that can be considered the new religion that deifies the state, characterized by nationalism and socialism or Marxist communism. Once again the world is returning to political barbarism, replacing the symbiosis "church and state" with "party and state," destroying the biggest political advances in the West, a representative democracy and the liberal state.

Biology is the nearest basic science to the social ideologies and in turn has greater influence on popular beliefs. The election of a random, completely fortuitous principle, as the origin and development of life, is equal to giving up totally on finding a theory for biology. In the same way, the hypothesis of spiritual entelechy or vital force that makes necessary the evolution of the universe toward the phenomenon of life puts an end to all progress of the scientific theories in biology. The connotations of one side or the other are not exactly beneficial to the search of knowledge.

New horizons are opening up for theoretical biology and for knowledge in general, following the successful road of Western science that in its rise avoids dogmatic temptations with "scientific" appearances. Great experimental advances in contemporary biology don't only possess an extraordinary potential to advance well-being for mankind, but rather—together with new developments in mathematics and physical chemistry—are forging new tendencies that carry the stamp of uncertainties that modify the determinism of physics in general; they are not restricted to classical mechanics, but simply refer to the complexity in themselves that represents reality, independent of interpretations in the macroscopic environment or atomic dimensions. Prigogine and other investigators have suggested modifications in the formulation of quantum mechanics in accordance with these tendencies.

It is considered that in the evolution of the universe, the first principles of physics have given us a coherent theory of its history, the theory of the big bang that describes mathematically, from the formation of subatomic particles to big material conglomerates, the galaxies that include stars and planets. Is this history necessary? Is this the only one possible? Science doesn't solve this unknown, nor can it. The same principle of objectivity on which it is based, only requires scientific hypotheses to be consistent and compatible with observation; it reduces scientific knowledge to an explanation or an ordering of the natural phenomena inside the structure of space-time. In addition, within the same physical principles, other universes would be possible by merely changing the magnitude of some constants or universal mathematical parameters. Chance and necessity are always present in the universe of contemporary physics that postulates the theory of everything, using the same principles for the same atoms and the same particles and waves in all places and times.

Nevertheless, the events we call biosphere are the limit to where our knowledge reaches; solely, we only know that they happen inside our home, planet Earth. In that duality of the world of physics, chance and necessity, is there only one region where the almost-impossible thing prevailed, biosphere and thought? From the point of view of objectivity, only if we discover life in other regions of space-time, or artificially replicate life and its evolution, will we be able to answer questions regarding determinism and randomness in the game of nature.

At the present time, new physical and mathematical concepts prefer to present indetermination as a result of a dialect between chance and necessity. These are represented by means of mathematical methods that are numerical solutions of nonlinear equations that describe processes in which multiple complex interactions intervene, such as the calculations used in the predictions of meteorology that pioneered models that, using new computers, promised to find stable and defined solutions. The results obtained initially in these calculations proved to be disheartening; these calculations were very sensitive to small variations of the initial data imposed on the system, a physical-mathematical phenomenon called "the butterfly effect," that seemed to indicate the impossibility of prediction of these systems.

Incidentally, this result, which had already been stated theoretically by the famous mathematician Poincaré at the beginning of the twentieth century, was later analyzed by Edward Lorenz in his meteorological calculations at MIT, focusing on the mathematics of systems that never find a stationary state—systems that almost repeat into themselves, but never in absolute form. His publications joined these two fields, mathematical theory and modern computational methods; the famous equations that Lorenz published in 1963, represented in three dimensions in Figure 28, have applications in meteorology that show the relationship between the peculiarity of weather that doesn't repeat itself, and the inability of meteorologists to predict it, suggesting a connection between nonperiodicity and unpredictability. Beginning in 1983 with the rapid growth of high-speed modern computers, a new scientific and mathematical field emerged that has created a new form of understanding of the high degree of complexity in nature.

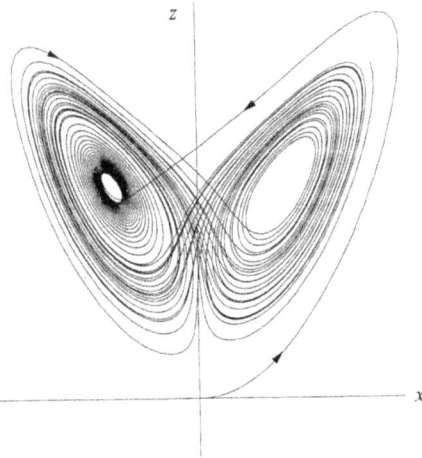

Figure 28: LORENZ'S ATTRACTOR IN TRIDIMENSIONAL SPACE OF PHASES

The laws of physics that are deterministic for simple mechanical systems lead to an indetermination in the global behavior of complex systems, which are products of mutual interactions of their components. The mathematical base of this theory of chaos is based on the calculations that show the influence of small fluctuations of the variables that define the system, being amplified to big interferences in its collective behavior that can be described by means of hidden mathematical regularities, within a nonperiodic evolutionary development, subject to bifurcations that indicate indetermination inside certain limits characterized by dynamic patterns. Figure 28 shows a point in the space of phases that moves parametrically with time in a random way with a basin shape around a center, to continue all of a sudden in a second similar basin, and so forth.

Philosophers of science, such as Stephen Kellert,[1] explore the interaction between methodology, epistemology, and metaphysics in the context of nonlinear dynamics and chaos theory. Although the so-called chaotic phenomena have been widely studied experimentally and accurately interpreted by highly sophisticated mathematical models[2] in computer technology, it is not usually considered that the implications of this theory could be strictly incorporated as a basis for theoretical physics. Even so, there is no doubt about its impact on the ongoing development of natural science, as well as the economic and social sciences.

Historical considerations provide understanding of this situation; the theory of chaos has generated an intellectual inertia that recalls the development of ideological breakthroughs, such as Newtonian and Quantum mechanics, which also passed through situations of discontent in the scientific community. On the other hand, it is interesting to note that the theory of Relativity with its rigorous logical foundation, based on more general principles than the heuristic method used in Quantum mechanics, was largely ignored until its spectacular first experimental test (proof performed by Sir Arthur Eddington in 1919). This proof made the front page of most major newspapers as well as Einstein as his theory of Relativity world-famous (see section 5-4-4).

Let us not be surprised, then, that distinguished scientists as Penrose and most of the admirers of quantum mechanics, have considered that chaos theory is nothing more than an empirical expression, limited to studying macroscopic phenomena of classical mechanics. Paradoxically,

---

[1]Stephen H. Kellert, *In the Wake of Chaos* (Chicago: University of Chicago Press, 1993).
[2]Steven H. Strogatz, *Nonlinear Dynamics and Chaos: With Applications to Physics, Biology, Chemistry, and Engineering* (New York: Perseus Books Group, 1994).

this cautious attitude is not consistent with the positivistic interpretation of quantum physics, (see meaning of the wave function, Section 5-6-2 e) which states that some of its fundamental rules have no explanation and should be taken as such, regardless of the interpretations some will wish to make in a philosophical context. Similarly, we could consider the validity of the mathematical formulations and methods of chaos theory, unrelated to philosophical interpretations.

However, there are serious scientific approaches, as expressed by Ilya Prigogine,[1] in the scope of thermodynamic systems far from equilibrium conditions, which are applicable to many experiments, such as the well-known "chemical clocks" that are visualized by color changes that aperiodically alternate, simultaneously resulting in blue- and orange-colored waves. Such systems, designated as dissipative structures, can allow the flow of matter and energy to build functional and structural mechanisms characterized by complex nonlinear interaction, leading to spontaneous self-organization. In contrast, in open systems close to equilibrium the second law of thermodynamics dominates. It inexorably produces a maximum of entropy, which implies the most probable state, and maximum disorder.

In a similar manner, the biological systems of a cell are perceived as dissipative structures. Some of the complex sequences that occur in biochemical reactions have been studied in detail, so that their characteristics can be assimilated to mathematical algorithms implemented for less complex systems such as chemical clocks. These models lead us to the simulation through software of phenomena that can be compared to that which made possible the initiation of life.

From all these philosophical, mathematical, and scientific ideas, principles have emerged in the new biology, characterized by key phrases such as "self-organization" and "emergent processes," which reveal surprising hidden laws of nature. How do these new laws work? Are they fundamentally different from the reductionist laws or linear laws? Stuart Kauffman[2] contends that complexity triggers self-organization, in order to overcome a certain threshold that favors the "emergency" of a new order, which he describes by the term "order from free," a very typical Anglo-Saxon expression, which could be interpreted as "not to owe anyone."

---

[1] Ilya Prigogine, ¿Tan sólo una ilusión? Una exploración del caos al orden (Tusquets editores, 1997).
[2] Stuart Kauffman, At Home in the Universe: The Search for the Laws of Self-Organization and Complexity (New York: Oxford University Press, 1995).

Specifically, when a sufficient number of molecules reaches a critical threshold of complexity and diversity, the system initiates an "ongoing process" that can lead to a new dissipative unit, a living cell. Thus, life is highly improbable, but almost inevitable.

Kauffman presents these extraordinary principles in the context of simplicity. He uses an analogy of thousands of buttons that are placed on a carpet, randomly joining them by wires. Initially, we have only isolated pairs; later on small groups of joined buttons show up, and suddenly, when about 50 percent of the total has been joined, a continuous or giant network structure emerges. In science, we speak of "phase transitions" in the initiation of processes, such as rain or the formation of small crystals of snow that suddenly appear, as might happen in the beginning of life.

This simple model shows the formation of emergent structures, which can be described mathematically by the laws of probability, using an algorithm or mathematical procedure that we call a computer program. So, computers have simulated phenomena such as the synthesis of sets of molecules, such as proteins, from a given system, defining possible reaction mechanisms leading to the system's self-organization. Let us express these ideas in Kauffman's own words: "Biological evolution may have been shaped by more than just natural selection. Computer models suggest that certain complex systems tend toward self-organization."[1]

The successes of molecular biology, genetics, and neurology must not hide their local value that requires new theories in a broader context within science. Moreover, in the inner world of our mind, physicists seek reformulations of quantum mechanics to consolidate the non locality or correlation of events outside the space-time, four-dimensional theory of Einstein, so that mental phenomena such as consciousness could be brought into science.[2] The concepts of necessity and chance are intertwined as the Buddhists' extremes of the yin and yang.

E. REDUCTIONISM AND HOLISM

To come closer to an understanding of the evolution of life and its origin is undoubtedly the most appropriate road for the formulation of a theoretical biology. The fundamental ideas of the evolution of life are derived mainly from observations of the environment bounded to our planet, interpreted by means of multidisciplinary scientific methods that

---

[1] Stuart Kauffman, "Antichaos and Adaptation," *Scientific American* (August 1991).
[2] Amit Goswani et al., *The Self-Aware Universe* (New York: Penguin Putnam, 1995).

range from geology to physics, passing by biology and chemistry. Their description is exciting; it is the closest history of our own existence that has a decisive impact on our beliefs of all order. Biology's individual and social influence is still in the developmental stages; it has an extraordinary potential to forge the next stages of humanity's development, as much in its physical well-being as in its intellectual development.

However, it is necessary to recognize that in contrast with physics and chemistry, biology still lacks a firm theoretical base, regardless of the big experimental advances that have been achieved. What scientific aspects should constitute that theoretical base? Are these principles the same ones used by physics and chemistry? It has been considered traditionally that the complex biochemical mechanisms that occur in the transformations of microorganisms, as much in their metabolism as in their reproduction, as well as the physical-chemical structural description of the macromolecules involved in living beings, is all that science can contribute in the study of the phenomenon of life. If this holds true, we are attending the end of the theory of natural science, just as is said to be occurring in physics. On the contrary, new concepts are opening up a road in the search for a biological theory that contributes new horizons to the current physical-chemical theories.

One of the general concepts that is considered fundamental for the new science is described semantically in the duality of reductionism and holism. In relation to the evolution of life, it is of great importance to consider that the simplification "cause and effect," which in linear form has remained current in Darwinian theory, is inadequate mainly because it only considers the effect of the environment on the subject and not the inverse, of life on its environment. At the present time it is widely considered as established biological knowledge that microorganisms that evolved in the first two billion years of life on Earth exerted a radical transformation on the terrestrial atmosphere, as they changed their primitive metabolism based on energy production by utilizing hydrogen, and generating oxygen later used by themselves. In other words, the primitive microorganisms changed their own environment gradually, and then adapted to that new situation by means of genetic changes.

These facts and many other similar ones in the history of life indicate to us clearly the reciprocity or circularity between organism and environment; a subject and its environment cannot be considered but on the whole, with a dynamics that is denominated nonlinear. When considering the origin of life, we imagine that the first organism appeared or was formed in a certain environment and evolved by adapting to it.

There is no such first living organism, because it cannot be considered without its environment; both constitute a unit with all their complexity. Their evolution occurred simultaneously; it doesn't seem probable that science can find in the biosphere trustworthy markings of those pristine stages of life.

The route to follow in looking for possible prebiotic stages should be based on scientific ingenuity, practical and theoretical, that simulates possible primitive states that work to create artificial life, even though it is not properly the same one that we are part of. These experiences are already being explored in the laboratory, based on physical-chemical reactions in virtual form, by means of mathematical algorithms that evolve with modern computer technology.

One of the biggest barriers to experimentation in this field resides in the limitation imposed by time. If we are simulating in the laboratory possible physical-chemical transformations that drive toward life, we are working with very short time frames; our individual lives or even generations of life are so brief compared with the geologic periods that lapsed in the formation of the biosphere. These reflections make us understand the extraordinary experimental difficulties that exist in this type of investigation.

The search for theoretical foundations for biology is undoubtedly the necessary support to facilitate the task of eventually devising experimental conditions that approach the primitive stages that gave place to life.

PART *Three*

THE INTERIOR OR SPIRITUAL WORLD

6

MATTER AND SPIRIT

In the title of this chapter, the word *matter* is first only because alphabetical order has been followed. In translating this book into English, we chose to respect this order, rather than following its inverse, as it reads in the original Spanish (*Espíritu y Materia*), to avoid prejudices that could predispose the reader regarding the importance of the topics that these words raise.

6-1 LINGUISTIC ASPECTS

The influence that these words, *matter* and *spirit*, have had through the ages leads us to a hierarchical order that seems to have been directed in the sense of diminishing the preponderance of the spirit and awarding it to matter. What a storm of ideas these two words bring up, from which it is so difficult to extract ourselves. In our time we look with mistrust to the word *spirit* and prefer replacing it with *mind*, which has a scientific air to it, leaving the former to refer to something vague and diffuse that suggests humorous anecdotes or feelings of fear that remind us of the afterlife or death.

What a big difference, if instead of considering these words as nouns, we move to the adjectives, *spiritual* and *material*. In doing so, we pass to another universe; such is the power of language. Years ago the international press published the incredible news that Pope John Paul II didn't believe in hell. In a continuous and unparsed article, it was explained that in a dialogue with journalists, the pope stated that "hell is a spiritual place," and therefore so was heaven. There was no known further controversy, because for the journalists, their own conclusion was definitive: A place cannot be spiritual. Then, must it be material? Yes, as it is affirmed by the modern judges of the truth.

On the other hand, it is commonplace to state that our modern society is "materialistic," as opposed to the past, which was characterized as "spiritualistic." A political or religious tint is frequently given to these concepts with a strong ethical character. The so-called "materialists" prefer to avoid employing the word *spiritual,* which sounds religious, instead replacing it with other phrases that have a scientific air to them: "human values," "humanistic concepts," and "humanitarian help." In other ways, the same word has an ethical characteristic, as when it is said that somebody is "very human" or we should "humanize the war," as if war existed without humans.

All these characteristics of language lead us again to value words as the symbols they are and to understand that their nature is conventional; that is to say, they refer to an agreement or covenant in a direct or tacit fashion that makes communication possible. Not to realize fully that words, according to the context in which they are placed, can have very different meanings and a limited validity within a historical and geographical environment may create great confusion.

Today, at least in Westernized societies, our children are taught that everything is made of or constituted by matter. I don't believe that teachers analyze deeply what they are teaching, even though it is possible that children may do so, although not in a very conscious way. They are being taught that all that stimulates our senses is called matter, including ourselves. This doctrine may be supplemented in science class in a more defined manner, by explaining that matter is composed of elements or different types of atoms, which are diffused in larger or smaller proportions throughout all bodies, and that modern physics has demonstrated that these atoms are formed by aggregates of other fundamental particles. What are these other particles? This question may be posed by an adolescent who has not had his or her restless "spirit" erased; the teacher responds to him or her that they are also matter.

The philosophy professor, addressing the student unfortunate enough to receive those useless teachings, may offer the student an analysis similar to the linguistic focus that we will present here shortly, or he may guide the student in a more classic form toward more abstract concepts that are called metaphysics.

Employment of the word *human* in phrases like "humanize the war" implies a redundancy because we know that war is essentially human, and that other animals don't practice it, at least within the same species. This statement, which seems humorous at first, can make explicit the use of the word *human* to designate some characteristics of the species that are not common to others. In general, the plurality of "human" characteristics could be divided in two groups: one which would be tentatively common with other animal species and the other strictly human. Certainly, when making this differentiation, it will be convenient to restrict it to certain species of animals, such as mammals, as well as to some opposed and similar characteristics, such as physical qualities, or to individual or social behavior, extending our list indefinitely. Will there be a way of avoiding this multiple plurality, establishing without any doubt the duality we desire at all cost?

Language seeks this miracle and carries it out with words we call names or nouns that magically give us the satisfaction that we have left confusion and seen the light. We say that we have "materialized" our ideas and feel a great relief. The words are from this point of view a deceit. It is worth quoting Schopenhauer: "Thoughts die the moment they are embodied by words."[1]

Frequently, when asking somebody a question, as, for example, about a technological novelty, we are answered with a single word: It is called this or that, real, virtual, digital, and so forth. It is as if the person is saying, "Didn't you know that?"

The difference between a noun and a verb can be very subtle. Infants frequently learn verbs first, or they invent them before names, insinuating that the event is a more primitive idea than its assumed cause, the name. In English, as a general rule, gerunds are used as nouns. In languages closer to Latin, such as Spanish, this modality is not as common. The word *ser* in Spanish is a verb and a noun, and so is the word *being* in English. However, *estar* is different than *ser*, and in English they are designated by the same verb, *to be*. Verbs express the happening; subjects, the myth. Citing José Ortega y Gasset: "A man is not a thing, but a drama: his life, a pure universal event that happens to each one and that each one is not in turn more than an event."[2]

The famous sentence of Descartes, "I think, therefore I am,"simply suggests that the verbs *to think* and *to be* are equivalent to a noun, the *me*. [3]

With these ideas in mind, we can come closer to the symbols that awaken so much respect: spirit and matter. In English, the word *matter* is also used to refer to what is important; in consequence, *spirit* would be the opposite. *Matter* derives from the Latin *mater*, which in Spanish

---

[1] Russell, *The History of Western Philosophy* (see Chapter 4, Section 4-3).
[2] José Ortega y Gasset, *Leciones de Metafísica: Obras Completas* VI. (Madrid: Revista Occidente, 1982): p. 32.
[3] Manuel Cruz, *Filosofía Contemporánea* (Taurus Ed., 2002).

means "mother," which is to say it indicates cause. Matter is the mother or the cause of all things, according to the fundamental teaching of the scientific era.

Then what does the word *spirit* mean? By opposition, we will say that it is not *matter* and, in consequence, in science it doesn't mean anything; it is a relic of the past. If this position satisfies us, then we are erasing with one stroke the whole philosophical, religious, artistic, and literary tradition; they are victims of a word that has confined us to a narrow prison.

In my youth, I received a humanist formation, by family tradition that was reflected in my father's library, where little could be found of scientific character. The cultural environment of that time in Colombia was traditional, and the teachings emphasized literature and philosophy, which was marked by scholastics. Scientific teaching was a novelty that was presented with great enthusiasm as the factor that would allow us to enter into the modern era and progress. These ideas instilled in me a great admiration for science that would be fundamental in my formation and professional orientation.

This book would never have been written without the influence of the intellectual context of my childhood, which has returned to my memory as it usually does in the golden age. The "spirit" remained latent, capturing from its rigorous scientific formation its intellectual solidity, without forgetting that in humanism there is a duality of matter and spirit. With a pendulous movement we can oscillate between the two understandings, but gravity will always remind us of the way back when we polarize ourselves; it is the law of existence. To return to the philosophical jurisdiction of linguistic symbols, we should consider metaphysics.

Metaphysics is as old as the thought process is. What is there beyond the direct sensations we perceive? The easiest answer is to invent a word, a name. That is the cause, the essence, of the real world. We have discussed the meaning of cause and we will say that a symbol cannot be the cause of anything. Causes are indemonstrable relationships among phenomena. We only know that the relationship of cause and effect is a mental habit that gives us an expectation of probability. This idea doesn't agree very well with our search that desires to go beyond our direct perceptions and into metaphysics.

It seems that the better answer regarding metaphysics would be that we cannot go further than the events; it is a dimension we cannot imagine. Only words are left, which were invented for that reason. According to this linguistic, symbolic understanding, we arrive at a classification of events or happenings in the same way we can classify books in a library. We could decide that the classification was dual or binary, and to extend it to the infinite, following the progression in base 2, with n as the exponent that represents the number of times that we disposed of this formidable task that would begin with two words: *spiritual* and *material*.

Natural science can help us define these two classes: material refers to all the phenomena that we can understand or explain by means of the laws of physics, considering physics as all the natural sciences—that is to say, those that refer to the events perceived directly by our senses. Following the binary procedure of mathematics, spiritual will be everything that is not material. If we are fanatics of science, we will immediately state that there are no events inexplicable by physics, that natural science understands the whole of human knowledge and that the rest is wordiness—ghosts or spirits that will disappear as science progresses. The section that we had dedicated for the spirit will be empty, or it will only be used for books that are considered worthy of a historical museum.

Let's assume that there is a group of scientists who have certain influence over the organizers of the great library, and they propose that although we have erased the word *spirit* and confined it to the museum section, we should look in the matter section for another duality: mental and physical. As for the section of physics, we have already come to an agreement: Everything observable by our senses as physical and mental will be the opposite. Mathematical logic is a symbolism; it doesn't derive from observation, and so it is mental. Philosophy, according to a modern point of view, studies the meaning of language; it will also be mental. Now then, the natural science, physics, has a dual character in inductive logic; the classification won't be appropriate. Then let's define the mental section as everything that is private, or internal, and the physical section as all that is public, or external. We thus derive the duality of metaphysics and physics.

Are mental and metaphysical equivalent? It would seem that they are not: Mental refers to a region of space-time, the mind, and has a physical character; it is external. We can study it by observational methods. Then what is metaphysics? Is it a ghost, the ego, the spirit that refers to mankind, the soul that Plato proposes? Is it a characteristic of all living beings, the "animism" protested by biologists? Or is it maybe the universal idea whose existence Plato didn't doubt: mathematics, logic, and all the concepts we call spiritual?

When we finally understand, after "superhuman" efforts, a mathematical or logical theorem, or by an intellectual gleam a creative idea emerges that appears to illuminate everything, an extraordinary happiness takes possession of us. This experience can only be compared to the artist's enjoyment when capturing harmony or brilliant stridency in his work, or to the mystical ecstasy that transcends all experiences past their finiteness. In contrast, sensual experiences provide us with the enjoyment of our senses, the ultimate expression of which may be the sexual orgasm. That is the duality of our experience: Should we choose between them? Is either experience fundamental or are they inseparable?

Plato's ideas are conceived as preexistent to sensations, which is to say to ourselves, in philosophical language, they are designated *a priori* knowledge, a Latin term of equivalent meaning. It is thought that these "ideas" are within us, independent of our individual experiences. Mathematical and philosophical truths are eternal: They don't depend on our experience; they are not part of our memory; they transcend time and individuality just as musical, poetic, and pictorial beauty do.

The concept of the soul is also Platonic, and it united with the religious tradition of the Bible to form a great synthesis in Christianity. The Greek philosophical-spiritual understanding was an integral part of the Christian doctrine, which exists in our day to a lesser or greater degree according to the cultural evolution of the peoples and social strata referred to. In contrast, we can consider that Western science, in its struggle for knowledge derived directly from observation, heads toward an understanding where the physical character prevails, directly inferred from data perceived by our senses. From this perspective, the "scientific" point of view has an extremist and excluding tint, leading us to a belligerent character, one of competition, should we transfer it to the ethical world of our desires that, in our index of the great library of knowledge, would be classified as politics or religion.

In this ethical analysis, if we desired to stay within intellectual justness, it would be preferable to differentiate, at least for discussion purposes, the existential aspects that determine our elementary desires from the philosophical ones that oscillate between these two tendencies of knowledge. Certainly the conclusion, if there were one that could be called definitive, would have important consequences in social as well as individual organization and development. In a certain way, we would be transferring the scientific method of natural sciences, based on induction and logical deduction, to the search for ethical and political values that could guide us in harmonizing our elementary desires with the group of Platonic ideas that we consider the basis of humanism—that is to say, those aspects of knowledge that have differentiated the human species in its evolutionary ascent. We should also keep in mind the inherent limitations of the inductive or scientific method, when it is applied to philosophical environments where it is not applicable in a rigorous way, as in mathematical physics.

As we move away from the mathematical environment, even within the natural sciences, and with more reason into the so-called social sciences, conclusions should be valued in a less conclusive fashion and certainly in no case in a dogmatic form that denies the essence of scientific thought.

6-2 METAPHYSICS AND SCIENCE

It is worth observing that within the natural sciences, the thought process of the great physicists is much closer to a philosophical and spiritualistic conception than that of the biologists, who are characterized generally as materialistic, when assigning to the word *matter* a reality that it doesn't have, transferring in subliminal form to their works aspects of the fight between traditional religion and science that have already been overcome in other arenas. Certainly, on a larger scale in the so-called "political sciences," we are in a much more favorable position to find conclusions and "scientific theories" that still have less credibility because they are to a larger degree influenced by our elementary desires and by the struggles of daily life.

It has become habitual today to consider as irrefutable the belief that matter is an absolutely clear and defined concept that definitively replaced the position enjoyed by the spirit historically. Biologists like Jacques Monod, in his book *Chance and Necessity*,[1] invoke the so-called objectivity principle that is equivalent to the process of scientific induction, giving it an absolute value that doesn't seem to be justified, not even when it is backed up by mathematical logic. According to these scientific dogmas, the study of the mechanisms of biochemical reactions—especially those interacting with DNA and RNA molecules in the synthesis of proteins, representing the basis for modern genetics—should lead us, even though they are still not totally "proven," to reduce all phenomena related to live organisms to biochemical mechanisms.

In contrast with biological reactions, chemical reactions generally refer to molecules with a reduced number of atoms in the study of classic kinetics in chemistry. In spite of the apparent simplicity of these chemical systems "without life," the mechanisms of the proposed reactions

---

[1]See Section 5-7-3 D.

indicate to us possibilities and not certainties that with highly simplified mathematical characteristics can be verified experimentally.

Ilya Prigogine, in his studies of physical-chemical phenomena, such as the so-called chemical clocks[1] and Bernard's instability, which are characterized by singular space-time auto-organizations visualized by colors and geometric distributions, has proven that they can be described by means of auto-organization mechanisms that are present under conditions far removed from those found in thermodynamic equilibrium. They suggest transitions whose mathematical modality cannot be described as deterministic, but by means of "patterns of distribution" in the so-called space of phases used in thermodynamics and other disciplines to represent systems with a high number of variables. The mathematics used to "simulate" these systems doesn't pretend to be deterministic, as the classic physical-mathematical style; on the contrary, it renounces determinism, though not to the way of the statistical mechanics of Boltzmann, which uses probability as the basis for its calculations.

This new scientific approach, which may be designated "organized indetermination," reveals a new concept of order that doesn't have as its basis random events governed by chance, which is the mathematical principle for statistics that is considered opposed to necessity. Again, the mental habit of extremes leads us to incorrect judgments. The necessity of the materialistic biologists represents the teleological interpretation of the spiritualists, who conceive a God who directs nature's evolution. On the other hand, chance is a consequence of matter that only obeys the laws of statistics, especially in certain cases like biological evolution. In these extreme and simplified stands lie psychological motivations that can be classified as neurotic cases that are expressed in the domains of religion and politics, transporting them to the land of natural sciences with the hope of justifying "scientifically" their phobias.

It is characteristic of the human spirit to admire that which we don't know with certainty, but which we intuitively sense by its appearance as spectacular. In our times, such adoration exists for the concept of matter, in a way that the readers of the Bible could compare to the idols of the heathens or the gods of the pagans. It is believed firmly that matter is a scientific concept; that is to say, that its justification is based on the mathematical theory of physics, and that therefore it is the universal cause. As physical theory evolves, its fundamental concepts decrease in number, following the so-called "Occam's Razor," according to which

---

[1]See Section 5-7-3 D.

all unnecessary principles are suppressed, reducing their number to a minimum. Well then, matter isn't strictly the basis for physical concepts; there is a preference to refer everything to "events" or happenings. Matter is a useful word for communication, but considered as the cause of all observable events, it is not a necessary concept for theoretical physics, such as force isn't in mechanics either.

Atomic theory arose from the philosophical concept of the Greeks of the indivisible thing, the atom. It evolved into the development of physical chemistry, from the primordial elements of the alchemists—air, fire, earth, and water—to the pure compound, the molecule, and elements of chemistry; moving on to the Rutherford and Bohr atom, with electrons and nuclei; progressing to the nuclear particles, protons and neutrons; arriving at quantum mechanics' countless families of particles, whose properties are determined by the energy levels used experimentally to detect them; until finally reaching the quarks, theoretical entities with occult mathematical characteristics that so far have not been detected.[1] If these quarks, products of mathematical logic, would be observed in some supercollider, physicists would have to complete their theories with other mathematical entities that gave them a logical base. Then, where is the matter? What is it? From the scientific point of view, we have reduced it to mathematical formulations.

In the same way, the physical basis of experiments and measurements has been evolving, from measurement systems such as CGS—centimeter, gram, second—closer to the natural units of our habitual world, to the natural units based on the constants $h = 1$ of Planck; $c = 1$, the speed of electromagnetic waves in a vacuum; and a third that could be $G = 1$, Newton's gravitational constant, whose fundamental nature is not considered so solid because of the difficulties that exist in reconciling gravity with quantum theory, leaving us in suspense when choosing it as a universal constant. The following dimensional relationships bring Planck units[2] closer to our habitual units of measurement:

Planck time: $(Ghc-5)^{1/2} = 5.389 \times 10^{-44}$ seconds

Planck length: $(Ghc-3)^{1/2} = 1.615 \times 10^{-35}$ meters

Planck mass: $(hc/G)^{1/2} = 2.176 \times 10^{-8}$ kilograms

Of these three units, we find mass to be the closest to matter, because it not only can be seen, but it can be touched; perhaps for this reason,

---

[1] See Section 5-6-2 D.

[2] Ian D. Lawrie, *A Unified Grand Tour of Theoretical Physics* (Bristol, UK: Adam Hilger, 1989).

before the scientific era, it was believed that the air didn't have weight and therefore was related to the spirit. After the theory of relativity, mass became equivalent to energy, which is more "ethereal," and time was equaled with space. What is there, then, after physics? Must we return to philosophy, to metaphysics, to Plato's eternal ideas, to the human mind that invented mathematics—or, if it pleases us, to the spirit?

The position of the so-called solipsism is well-known in metaphysics. This plausible hypothesis consists of assuming that the individual mind or the individual spirit is the only entity in which we can have faith, because the whole sensorial universe is inferred; it is a product of our mind. Therefore, this premise can take us logically to deny the external world and to conceive it as part of the individual spirit that Plato called the "soul."

Strictly speaking, this stand is impregnable; we cannot refute it, but "common sense" indicates to us that it is not applicable to daily life. It would lead us to an untenable situation: My world would be as the virtual world of the computer. If we were to adopt strict solipsism, the dangers we would confront would soon lead us to extinction, which would also be virtual and part of the external world, negating and affirming itself. Looking for the humorous side, if the conscious supercomputer of the future ends up being invented, perhaps it would genuinely be solipsist: unafraid of death in spite of possessing self-consciousness and the most powerful of all consciousness: solely solipsist.

Interpreting Kant, we would say that practical reason claims its validity, making pure reason possible. Nobody can live being a solipsist—maybe not even the autistic; there is no drug that can take us permanently to that pedestal. Perhaps the most ferocious rulers have been able to come closer to those heights.

From a positivist point of view, thoughts have their origin in the sensorial data that determine our individuality as it develops progressively, taking as a basis the biological inheritance that is preexistent to the individual, which at the same time comes from universal evolution. This general evolution of the universe, described by physics, implies an order in nature that also encompasses biological evolution and is neither necessary nor teleological, nor is it simply the product of chance.

In classical physics, where systems are idealized and their parts considered independently according to the Cartesian method, today called reductionism, there is a tendency to lean toward determinism in mechanics, even in cases where complex systems are considered, as in the statistical mechanics of Boltzmann. In quantum mechanics, the uncertainty principle is conditional to observation; in that sense it is solipsist: It doesn't have validity without the observer's intervention. The uncertainty is made reality when measuring a physical variable; it is not intrinsic to nature. Following a similar approach, we could think that an object ceases to exist when we are not observing it.

In cosmic physics, when considering the origin of the universe, its deterministic conception allows for the election of the initial conditions, microseconds after the initial singularity, including the universal constants, h, c, k, whose variations would lead to multiple possibilities in the global evolution of the universe. In local aspects of this evolution, like in the possible formation of the terrestrial biosphere, these physical theories may indicate the need to choose certain initial conditions for their development that may be considered consequences of the anthropic principle. However, the current formulations of mechanics are not adapted to analyze dynamic systems of great complexity as the phase transitions that should have occurred to enable the development of systems that are far from thermodynamic balance, such as the biosphere, which doesn't follow deterministic patterns. Is it necessary to wonder, if some fundamental discontinuity exists in the transition that allowed for the development of life?

Simplistic assumptions can take us to ideological extremes: The spiritualists, or animists, as their opponents prefer to call them, give reality to the word *spirit* to designate entities or causes independent from matter that is, in turn, the entelechy of the materialists. For those who are considered scientists, and therefore owners of the truth, there is but one cause for everything: matter. Those on the spiritual side, in a defensive position, prefer to favor causal dualism. The two positions are sterile as to their contribution to knowledge; the names are only symbols for communicating and as such have mainly social value. Eventually they may lead us to endless discussions with catastrophic, warlike effects or, more benevolently, to humorous stories that reflect a more "humanist" or "spiritual" stand, whichever we may desire.

A closer position to science will take us to the world of biology, in connection with the evolution of the biosphere—in other words, to its history. In the biological microcosmos there are unknown territories between the modern microorganisms, the ones in existence today, and some that are supposed to have existed in the chain of evolution that didn't leave any trace or fossils that allow us to objectively disregard that experimental discontinuity. In the natural sciences, it is possible to appeal to experimentation under laboratory conditions, in which the parameter time can be decreased to almost infinitesimal values—this not being the case in the opposite scale when we come closer to periods of time that, compared to our existence, seem infinite.

In the world of physical chemistry, today we dispose of observational means that go to the confines of the universe and allow us to receive spectroscopic information—that is to say, through electromagnetic and gravitational waves from all directions of space-time. This information is the physical-chemical history of the universe; we observe the celestial bodies in multiple stages of their development, from the big bang, traveling by quasars and galaxies arriving at our home, the solar system. The modern physical theories, quantum and relativistic, are consistent with that experimental and universal panorama. That this is not the case with the immediate environment of the biosphere, which seems to come from the universal evolution, is a startling occurrence. The experimental possibilities in biology, which expand in space-time as those in physics, are still very limited in the new spatial technology and are reduced so far to trips comparable to those of Jules Verne in the nineteenth century.

We then should remain for the time being within the confines of biochemical experimentation of local character. Just as chemistry, in a macroscopic experimental field, established the fundamental ideas of atomic theory that made possible the theoretical-experimental development of quantum physics, modern molecular biology with its extraordinary achievements is propitiating a new scientific focus that points to restating classical and quantum mechanics, making it accessible to systems of great complexity that are considered in a systemic form by means of new mathematical developments.

Prigogine observes, with great intuition, that in the phenomenon of life and particularly in the interface that gave its origin, a new physical understanding could be hidden that takes into account the irreversibility of time, its arrow. The fundamental ideas presented in his book *The End of Certainties* refer to the concept of "organizational patterns" of the so-called dissipative structures and to the introduction of an indetermination in the formulations of mechanics that interprets the transformations of complex systems of many particles, under conditions far from thermodynamic balance. The dissipative structures are models of dynamic systems that can be compared to continuous-path chemical reactors, familiar to chemical engineers, where dynamic conditions of flow of energy and matter with the exterior sustain a "static dynamism" in the reactor.

Similarly, a cell, a microorganism, maintains its static vital macroscopic conditions inside a dynamism, or organizational pattern that implies nondeterministic fluctuations. In a more fundamental theoretical environment, the new mechanical formulations head toward a systemic understanding that analyzes the movements of a group of particles by means of the parametric variations of the distribution densities, corresponding to small fluctuations of the initial conditions that are amplified at a global level.

In this mechanics, indetermination exists at an individual level of the particle and therefore the unique trajectories in the space of phases have no meaning; it is said that a diffusion process occurs. The system is not deterministic and can only be described as a whole, and not as an average of individual trajectories. The statistical mechanics of Boltzmann, applicable under thermodynamic equilibrium conditions, is replaced by a new mechanics that refers to systems far from equilibrium that are fundamental in nature and that give continuity to the general evolution of the universe, including the formation of the terrestrial biosphere and other similar ones that could possibly be discovered.

From a philosophical focus, we consider that those new scientific concepts unify nature in its evolution and favor the understanding of a metaphysical monism as an underlying reality to the events that stimulate our senses. This conclusion makes us remember other entelechies of the past that have disappeared in the history of science. One of those words, the *phlogiston*, which is no longer in dictionaries, was assumed to be a fluid that moved under the effect of a temperature gradient. The experiments of James Prescott Joule on the equivalence of mechanical energy and heat, as well as the kinetic theory of gases, definitively identified heat as a form of kinetic energy, making useless the phlogiston. In a similar fashion, even though not in a scientific way, the word *apparition* had similar luck and was relegated to a folkloric vocabulary in the stories of encounters with the unknown that became less and less frequent with the use of the electric light.

If we continue on the topic of the influence of science on the evolution of language, we could refer to the famous "ether," assumed to be the material support for the electromagnetic waves, which disappeared with the quantum theory that justifies the wave-particle duality mathematically, as fundamental in nature. It would be expected that this process would continue with words like *spirit, soul, god,* and—why not?—*matter.* In this field, the struggle seems to be much more difficult; if one takes into account that these words raise feelings that move us, they belong to the realm that relates to individual and social human behavior.

6-3 EPILOGUE

A complementary point of view considers strictly metaphysics and refers its concepts to the science of the mind, passing to the world of psychology. We have had a chance to analyze some of these issues, relating them with the acquisition of knowledge and with regard to the development of the concept of ego; later we will study their importance and projection in mysticism.

As a matter for discussion let's consider metaphysics in a more independent and therefore closer way to its historical origin: Let's give reality to words, return to our starting position and refer ourselves to the duality of mind and matter, to be more specific. Common sense tells us that there is something real that corresponds to those words; we assume that it is this way. What can we affirm or define regarding these assumed entities? Matter is the underlying reality, the cause of the whole observable by our senses, including our body, our brain. In this sense, matter would be the origin of all events and there would be nothing more to speak of. Physics, in its wider sense, studies all these phenomena and relates them in a logical and coherent form. The mind is formed by the non-inferred data, or impressions imprinted by matter on our brains, which are also matter. Are these data equivalent to those that are registered by a photographic camera, or an instrument able to gather all or more than the capabilities of the senses?

We believe that the mind has the ability to elaborate and relate this data, creating a complex web of communication. Well, then, we can say that a computer does something similar, even though the latest state-of-the-art model is not equivalent to the mind; the possibility exists that it may appear, without ignoring in absolute form that there are valid theoretical reasons to which we have referred previously, that put in question that possibility. With these exceptions, let us suppose that said supercomputer can be built, even one overcoming the human mind.

Nature in its evolution has arrived at the human mind, and in a more advanced step to that supercomputer; on the other hand, the possibility of other minds, products of the same material evolution, which have evolved in other places of the universe exists. From the point of view of this analysis of the entities known as matter and mind, we will have ended up affirming that they have the same cause. Now then, knowledge is established by noticing diversity. Is it necessary to ask if the concepts of mind and matter are or are not identical? Later Wittgenstein, in his idiomatic disquisitions, so admired in modern philosophy, talks to us about "the composite and the simple,"[1] and asks: How is it that names actually designate what is simple? What are the constituent parts of which reality is composed?

All these questions and indefinite answers of later Wittgenstein can be summarized in a compact way, to the way of earlier Wittgenstein (*Tractatus*). Good reasons adduce Bertrand Russell[2] to move away from the later Wittgenstein, as he commented on his digressions. In my opinion, knowledge derives from noticing diversity, and language symbolizes that diversity, assigning parts by means of words. The limit of this game is in our capacity to differentiate any event from another. This mental capacity generates thought and enriches language, which evolves in parallel.

Returning to our topic, the fundamental part of the analysis of the concepts of matter and mind is to distinguish them; it is to establish their diversity, and it doesn't matter what their "causes" are one or the other. In a similar fashion, the same way we can end up conceiving that the cause of the whole universe is matter, it is licit to state that the "cause" of everything is also the mind. If we consider it singular, we will become solipsists and if we find that it is collective and we stay in the human environment, we come closer to the psychology of the "collective unconscious" of Jung, and progressively we will be able to pass to the universal order; to Plato's ideas or mysticism.

The spirit or the universal mind, whichever we may prefer, can be considered "logically" as the cause of everything; it suffices to consider that knowledge is not possible without the mind. Consequently, there is neither now nor here, neither space-time nor physics that studies matter, without the mind, without the spirit. If we say in a very scientific fashion that the mind is a region of space-time or organized matter, capable of receiving and interpreting data, we will also be able to argue that matter is an invention of the mind. How may we solve this seemingly insoluble dilemma? We will say then that the duality of mind and matter constitutes two aspects of existence that we fortunately distinguish and that when separated lose meaning.

In a similar way, quantum physics arrived at a dual understanding of an event, by means of the interpretation of the wave function, a mathematical artifice that characterizes a mechanical system—for example, a free particle that moves with a given speed. This interpretation describes its position, with wave-like properties in space-time, and in turn codes information on other physical variables such as its speed and energy.

---

[1]Wittgenstein, *Philosophical Investigations* (see Chapter 3, Section 3-3).
[2]C. W. Kilmister, *Russell* (Hampshire, UK: Palgrave Macmillan, 1992).

This theory establishes a duality of nature, giving to the material particles, such as a ray of neutrons that fundamentally is characterized as such, a wave-like character, and to the waves, such as a beam of light that is basically considered of electromagnetic wave nature, the characteristics of particles.

Could it be that mathematical physics brings us closer to an idealistic understanding of the universe? Is there a tendency in that direction in modern physics, with renowned scientists such as Heisenberg, Bohm, Prigogine, Penrose, and Capra, among others, presenting in famous works of scientific divulgation measured points of view that move away from contemporary science, from the materialistic paradigm that threatens to reduce modern thinking to narrow conceptual margins? New horizons point to undetermined formulations that don't justify the unalterable laws of nature, nor are they based on the evolution founded in randomness. Contemporary Western philosophy, on the contrary, oscillates between linguistic indetermination, and its disguised admiration of science, and its method based on mathematical logic. In contrast, the old understanding of Eastern philosophy is being popularized in the scientific environment, from physics to psychology.

7

TRANSCENDENCY OF THE EGO
AND MYSTICISM

In the dawn of human evolution simultaneously emerged its consciousness, the ego, and mythology that transcend individuality, which
is projected to the external world in the form of spirits and beings that
idealize it. In the history of each human being, the project of life in its
biological and spiritual form, molded by the physical and cultural environment in which it is developed, repeats itself like in a hologram.

Each individual mind is an indispensable factor in the collective survival that is present in a larger or smaller degree in all living beings, in such
a way that it can be considered as a consequence of the emergent universal
order. Without this feeling of individuality, of being different and unique,
we would not have the element we call consciousness, an indispensable
factor in the development of life, at least in the species we call superior. We
human beings take our self-centeredness to such a point, that we consider
individual and collective consciousness to be unique to our own species,
even though exceptionally different criteria have coexisted in this respect.

7-1 FUNDAMENTAL RELIGIOUS BASINS

The great cultural traditions we designate as Eastern and Western in
simplified form, have very dissimilar understandings regarding man's
position in the context of nature; their paradigms are an extension of
the feeling that individuality plays in connection with the external world.
Their conceptions of the ego are reflected in their mythology: The west
imagines its God, as one and only, personal, as each one would want to
be, as the sublimation of the ego, all mighty, powerful, kind and with all
the attributes desired for himself. In the Catholic Spanish version of the

Bible these two aspects are explained clearly: God created man in his own image and, in an apparent contradiction, the biggest sin for a man is his pride, his desire to be like Him.

In the East, the secular tradition doesn't exalt the ego; on the contrary, it recognizes its action that creates and destroys as a part of the eternal struggle between the yin and the yang, the poles of nature that harmonize by means of an evolution of the cosmos in which life, and in particular mankind, is not the center but a consequence of that progression. Eastern philosophy doesn't conceive a personal God that as a Father creates the world and directs it; nature in itself, with its diversity and order, is its own cause, which is expressed mythologically by the snake that devours itself and by the goddess with multiple arms.

These two major philosophical and mythological ideologies that have been developed in the "Old World" have not been isolated, nor are they monolithic. Their Euro-Asian geographical center evolved from a great diversity and pluralism toward those two directions that today we recognize and that hopefully will converge again. Arnold J. Toynbee, the great English historian, presents us with a panoramic vision for the emergence of great cultures: the millennial Egyptian culture and its interrelation with the Mesopotamian culture, are the developmental poles that gave place to the Mediterranean civilization that we consider as the consolidation of the so-called Western tradition. This tradition, from the mythological point of view, is synthesized in our time by Christianity and Islam, both monotheistic religions that have a lot more similarities than differences, although from an endogenous point of view we are not willing to accept. In the same way that in the West we consider superficially that Hinduism and Buddhism have very close philosophical traditions, equally in the East few basic differences are observed in the monotheistic religions.

If we observe the phenomenon of cultural globalization that is currently accelerating, it is impacting to see that these two big cultural ideologies with such dissimilar philosophical and religious understandings still subsist. From a social point of view, great ignorance and incomprehension exist in the West regarding Eastern tradition, perhaps because of the Western arrogance that has its origin in the success of science, which has given it technological and military superiority. Just as Toynbee predicts in his book, *The World and the West*, in that order, and particularly the East with its superior philosophical-religious wisdom, are obtaining cultural conquest more rapidly than the West. To what point is Western civilization, even at the level of its intellectual elite, conscious of this historical tendency?

In these considerations of religious traditions it is necessary to note that within their own environments, very diverse ideologies exist such as mysticism, mythology, theology, rituals, and ethics that shape their fundamentals. The word *religion*, of Latin origin, refers as such to behavioral norms. This linguistic fact seems to indicate that the fundamental religious characteristic is directed toward obtaining a certain social organization. It is to assume that theology, as a philosophical and intellectual support for religion, was added later to the historical road of its consolidation. This is not so in the development of mythology, which is more of a symbolism and therefore closer to the primitive origins of human thought.

We generally associate the word *myth* with fantastic stories describing the origins of civilization, especially when we attribute them to religions that are not from our own culture. In the West nobody takes seriously the gods of Greek mythology; we find naive the Islamic sensual paradise, and we make fun of the pudgy Buddha who contemplates himself in a lotus position. Within our own Christian culture we still argue the validity of worshiping images that materialize mythological symbols for God, such as the angels and saints so dear to Catholics and orthodox worshipers. The so-called Protestant churches, today simply called Christian, consider their opposites idolaters, thereby coming closer in this way to the Jewish understanding of the Bible.

Within this whole context, religious ritual also appears, even though seemingly it is not an essential doctrinal element; it carries out an important function that stimulates and aids mystical expressions that are only developed in occult groups that transcend religious organizations. Within this ritual we should include, in addition to prayer, religious carols, sacred music, moments of silence at churches and convents, occult practices such as meditation, and elaborated psychosomatic procedures such as yoga.

7-2 THE PERENNIAL PHILOSOPHY, TRANSCENDENT AND EMERGENT

Aldous Huxley, in his book *The Perennial Philosophy*,[1] presents universal mysticism as a religious sublimation that extends from the West, with the primitive Judaic communities that gave origin to Christianity in its "Jesus" doctrine continuing with the big mystics, including Teresa of Avila, Francisco of Assisi, Master Eckhart, and Mohammed to the distant Orient with Lao-Tse and Buddha. The fundamental characteristic of the mystic

---

[1]Aldous Huxley, *The Perennial Philosophy* (New York: HarperCollins, 2004).

tradition is its unity, within the great conceptual diversity of the differ-ent philosophical-religious ideologies from which it transcends. Modern tendency for mystic understanding is directed specifically toward East-ern philosophy, as indicated to us by religious thinkers like Alan Watts,[1] who, in excellent writings such as *The Philosophies of Asia*, compares the Christian tradition with Hinduism, Taoism, and Buddhism.

The etymology of the word *mystic* is derived from Greek and means silence. In the Eastern tradition the key to mystic thought is expressed in holism that indicates unity and that conjugates opposites harmoniously. Just as silence supports sound, and light originates from darkness, the universe comes from a vacuum, in a harmonic way where there is no cause or effect, but vibrations between being and not being.

To come closer to the mystery of life, let's meditate on death; in an attempt to look for the "self" and the "you," let's look for their origin in the boy's empty mind that laboriously imagines the path through nature with the illusion of individuality. What fraction of our individuality is as original and proper as we assume? Our ego is a dream or an illusion imposed on us by nature to give continuity to the stream of life, when passing to its negation, death. Without it life does not exist, just as a vacuum makes existence possible.

The ego allows for continuity of life, creates its path; it is fiction, a mask, one without which we cannot act in the game of existence. It os-cillates between appearing and disappearing, yes and no, day and night, silence and sound, past and future, and yin and yang. This game can become a drama, unless we wake up and realize that behind that mask is all of nature and that there is no "me."

Originally in Latin the word *person* designated a mask from which voice originated. The verb *sounds* (son) is transformed through (per) the individual, the mask. This mythical aspect of the ego is present in a more explicit form in Christian gospel; when referring to the Son of God, it states: "The Verbum [in Latin, God Father] transformed into man and inhabited among us." God the Father becomes a person; he acquires individuality in Jesus, enters the game of life, and transcends with his death and resurrection.

The Eastern paradigm frequently presented in the West as pantheist, as the understanding that the world and God are the same, is not correct. To understand the idea of Brahman or the oriental Tao, we should disas-sociate ourselves definitively from the Western concept of a personal

---

[1] Alan Watts, *The Philosophies of Asia* (North Clarendon, VT: Tuttle Publishing, 1995).

God—in other words, from the concept that identifies God as Creator of the Universe. Passing over this threshold will allow us to initiate our understanding of Eastern philosophy.

Far is the Brahman or the Tao from being a person. Only mystic sensation can bring us closer to its holistic understanding, from the Greek word *holo,* or unit, that by its own nature is inexpressible. Strictly we cannot symbolize it but by means of metaphors, allegories that poetically, musically, propitiate meditation that leaves the mind empty and brings us closer to the primitive state where there is only sensation, without an observer, that invention, the ego.

Nature is a project of itself; it evolves in a stationary fashion. It has a dynamic permanency, to the way of a stream that always has a surface with a peculiar form, and at the same time flows. The Hindu history doesn't have dates; time is cyclical and eternal. Time is not important; in this way neither is our supposed individuality that is a page without a number from a book without covers.

Our consciousness is a minimal part of our nature that is oceanic; the symbolism of the gods with multiple extremities reveals the unconscious aspects that constitute the essential foundation of life: not being prevails over being. Lao-Tse in the *Tao Te King,* according to the version of Roberto Pla,[1] expressed it poetically, this way:

The eternal Tao doesn't have a Name
When it begins to "be"
Is when the names arise

The names also end up being
and they may find out where it is necessary to make.
When one knows where it is necessary to stop
one is safe

That is why Tao in the world
can be compared
with the rivers that run to the sea,
that know where it is necessary to stop

Return is the Tao's movement,
Development is the Tao's consequence.

All things are born in being,
Being is born in the not being.

---

[1] Lao-Tse, *Tao Te King,* trans. Roberto Pla (Mexico City: Editorial Diana, 1980).

In the Hindu and Buddhist mythologies, perhaps in their most popular aspect, reincarnation or transmigration of souls is dreamed upon, raising their importance. This belief is closer to Western tradition, through Greek mythology that is more kindred with Eastern culture, even though it is considered a curse in the Judeo-Christian orthodoxy.

Belief in human immortality is present in all religions, except in the materialistic credos that paradoxically drive faithful believers to a frantic race against eternal time that is the God in which they firmly believe, and that inexorably alienates them from nature.

The materialistic credo can also drive adherents to mystic states, although fleeting ones, when adverse conditions make their most outstanding exponents falter. Some inspired writings attributed to the extraordinary imagination of Nobel laureate Gabriel García Márquez, in his "farewell letter,"[1] illustrate his transcendency:

> If for an instant God would forget that I am a marionette made from scraps and He gifted me with a piece of life, I would take advantage of that time as best as I could. Possibly I would not say all that I think, but definitively I would think all of what I said.
>
> I would sleep little and dream more, understanding that for every minute that we close our eyes, we lose sixty seconds of light. I would walk when everybody else stopped, and I would awaken when everybody else sleeps.
>
> So many things I have learned from you, men. . . . I have learned that everybody wants to live at the summit of the mountain, without knowing that true happiness lies in the way we go about climbing its steep slope.
>
> I have learned that when a newborn first squeezes his father's finger in his tiny fist, he has caught him forever.
>
> There are so many things I have been able to learn from you, but in reality they will not be useful, because when they place me inside that suitcase, unhappily I will be dying.

7-3 PSYCHIC STATES AND FUGACITY OF THE SELF

Mystic vision can be fleeting before descending into hell. The ascent in the scale of states of our consciousness may take us to the peak of Nirvana; to remain there it is necessary to awaken the dream of our desires, and to go

---

[1] In 2003, when it was wrongly rumored that Márquez was terminally ill, this "farewell" letter circulated over the Internet.

up the steep slope without intentionally wanting to arrive anywhere. Only in this way will we always be prepared to halt and return.

In Eastern philosophy, we can see that the metaphysical aspects are preponderant in Hinduism, which is fundamentalist, deriving toward Taoism and Buddhism, where an ethical character based in the analysis of the characteristics of the human mind prevails. For this reason, psychoanalysis in the West manifests a great likeness with Eastern ideology in our time, as Ken Wilber[1] expresses in *The Atman Project*, which implants the Western psychoanalytical tradition of evolution of the ego within the arc of individual existence that ascends in the stages of its evolution, to return in a regression to its origin.

In the West, the development of the personality and its influence on the mature mind that one investigates by means of psychoanalysis is emphasized, reviving the trajectory that flourishes from the singular and collective subconscious. That natural tendency of the ego, in its different physical and spiritual stages, designated as the external arc (see Figure 29) in *The Atman Project*, has been considered traditionally in the West as an end in itself, and at least in the world of psychology, it doesn't go farther. Within the materialistic doctrine, pseudo-science postulates a new Epicureanism; values don't exist but those that derive directly from the enjoyment of our immediate desires, which wisely even though inadvertently are recognized as the foundations of the personality.

Mysticism in general and, in a more explicit and conceptual way, Buddhism considers the stages of development of the personality in a dynamic fashion characterized by a series of psychic states, patterns of

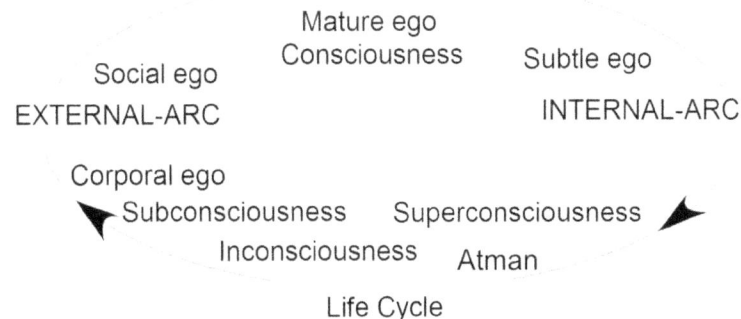

Figure 29: **LIFE CYCLE: THE ATMAN PROJECT**

[1]Ken Wilber, *The Atman Project: A Transpersonal View of Human Development* (Theosophical Publishing House, 1980).

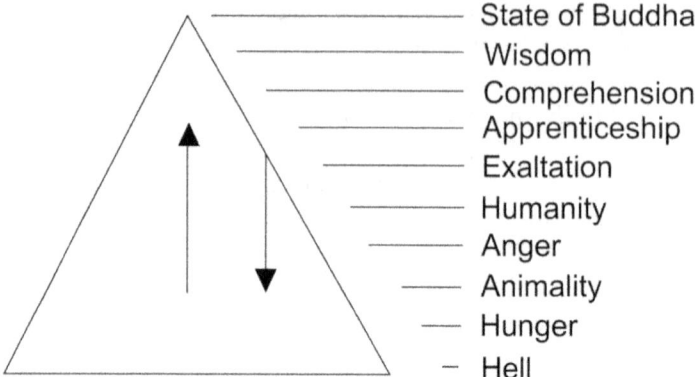

State of Buddha
Wisdom
Comprehension
Apprenticeship
Exaltation
Humanity
Anger
Animality
Hunger
Hell

Figure 30: TEN STATES OF THE WORLD: BUDDHIST VISION

mental evolution. These states of consciousness, according to Daisaku Ikeda[1] (see Figure 30), are configured in the human being by the feelings and desires that prevail according to external and internal circumstances, and determine basic emotions such as happiness, sadness, pain, fear, anger, and others.

These two approaches to studying the personality complement each other: The Western focus through psychology refers to the external arc of the evolutionary cycle, described by psychoanalysts and psychologists like Freud, Jung, Piaget, and Erich Fromm, among others; complementarily, the internal arc return in the cycle of life is conceptually mystic. The Buddhist vision has a pragmatic motivation—the search for happiness by means of analysis of desires and feelings that characterize our psychic states recognized as stages of development and evolution in an ethical dimension, which doesn't necessarily correspond to individual temporary evolution. These psychic states are not permanent, nor are they acquired like knowledge is; they have a dynamic nature that reminds us of mathematic, analytic, and graphic representations, described by nonlinear variations that implement concepts of feedback and self-regulation, key for the design of "cybernetic" models (in Greek, "pilot"), or steering of neural processes. These scientific aspects, actually called systemic, are presented masterfully by Capra in his extraordinary writings in *The Web of Life,* and have a calling to provide a great impact as a new path for acquisition of knowledge.

The human mind has two working mechanisms that have been recognized already in common language, fundamentally as conscious and sub-

[1]Daisaku Ikeda, *Life: An Enigma, a Precious Jewel* (Kodansha International, 1982).

conscious. In a broad view, we can consider within conscious "sequential intelligence," or logic, which allows us by means of deductive processes to use procedures that can be simulated by electronic computers, as well as processes of physiologic and psychic self-regulation that happen without the dominion of our attention. In both mechanisms, systems are recognized that include feedback and self-regulation cycles, characteristic of cybernetic development in its initial stage, which have given place in their frontier borders to consider self-organization as a phenomenon that drives hopefully to "intelligent machines" with the capacity for learning.

In what domain could we locate the mystic phenomenon, when relating it to cybernetic simulations? Undoubtedly, mysticism is closer to the subconscious that prevails in transcendental meditation, a method par excellence for reaching that spiritual state.

Complementarily, it is recognized in intellectual yoga and in Buddhism, the superior psychic states of wisdom and understanding, the bodhisattva that in Japanese designates wisdom and sensibility, which also imply altruism and compassion, the road toward the mystic experience, the nirvana. All these ideas point to considering mysticism as a global phenomenon of the mind, that in a similar way is also present in the deep understanding of the big thinkers.

Danah Zohar and Dr. Ian Marshall[1] define "spiritual intelligence" as spiritual characteristics of the same order that allow us to access creativity and ethical values that give a superior sense to existence. The modern neurological studies that try to elucidate cerebral mechanisms that happen simultaneously with certain psychic states—for example, the measurement of frequency and intensity of electromagnetic waves, as well as cerebral images that visualize areas where presumably a higher level of activity occurs—can be considered in their current state of knowledge as simple empirical data, until corresponding physical-mathematical theories are developed.

Frequently this physical data is presented in journalistic form, as explanatory of the spiritual phenomena: "spiritual transcendence is simply 40 Hz frequency waves in which certain cerebral regions intervene," insinuating boastfully with the verb *to be* that this is a great psychological discovery, which invalidates spiritual aspects such as mysticism.

This is not to diminish the importance of these neuro-physical studies, but rather to recognize their environmental validity. One of the qualita-

---

[1]Danah Zohar and Dr. Ian Marshall, *SQ: Spiritual Intelligence* (New York: Bloomsbury Publishing, 2000).

tive aspects suggested by these studies is the recognition that psychic states, such as the mystic phenomenon, the state of vigil consciousness, deep sleep, and the effects of stimulants and anesthetic drugs, can be compared by physical means as well as related to each other as global phenomena of the brain and perhaps of the total organism. This suggests they can be simulated by mathematical-systemic models, which, developed in cybernetics as well as in their conceptual foundations, are considered the frontier of science. In this way relationships are seemingly established between physics, in its broader sense, and the perennial philosophy, which is a historical constant that emerges as a characteristic of thought in all times and places.

7-4 ILLUSION OF THE EGO AND SPIRITUAL LIBERATION

One of those historical constants of the spiritual vision has as its foundation the oppressive reality that individual life, our ego, is an ephemeral phenomenon; it disappears with death. No being is indifferent to this reality, which is a future fact from the individual point of view and whose meaning to mankind is always present in our society and in the individual. Depending on the interpretation we give it, all of our existence depends on this phenomenon in great measure. We can avoid it, like in the modern society of materialistic tendency; worship it, as in the monotheistic traditions; or philosophically value it by means of mystic conception.

The materialist says good-bye to his friends and relatives, to the way of the comrades who usually say, without shedding a tear, the indefinite lapidary phrase: "until forever." In their places of work, the omnipresent loudspeakers hammer: "Get to work and produce" instead of "relax, have a rum." Television brings about indoctrination, especially in children, with *Pokémon, Power Rangers,* Nintendo, and others subjecting their parents and themselves to indefinite consumerism of all sorts of trinkets that quickly form part of the "cemeteries" that contaminate planet Earth.

In contrast, the Carthusian monks, that is, if there are still some around, die every day, every time they meet during their prayer walks: "Brother, of dying we have, the how and the when we do not know." The Clarisian nuns receive their novices at the entrance of their sepulchral convent with a great epitaph: "The pleasure to die without pain is worth the living without pleasure." The Buddhist monk, a refugee in Nepal, still misses the pompous ceremonies and the religious palaces that were destroyed by the "bellicose progressives" in the sadly celebrated cultural revolution that hopefully didn't destroy his transcendental spirit, which resurges to the way of the first Christians in the Roman circus, in a new mystic synthesis that points toward the four cardinal points, especially right from the heart of the Western Empire.

In recent time, the massive means of communication have brought to light some anecdotal stories implicated in the belief of the transmigration of souls, in reference to the Tibetan Buddhist traditions. Specifically, it is narrated how the monks choose the successor of the Dalai Lama, their spiritual leader, by means of criteria that seeks spiritual qualities that will define the boy into whom he has reincarnated. The Buddhists have elaborate theories regarding spiritual states that go beyond individual death, and possibly eventually restart new lives by means of reincarnation.

For argument's sake, to the Western way, we can assume that a spiritual continuity exists that goes beyond the evolution of nature as a whole, and has a reality in the permanency of individual characteristics that are not related by genetics, which is a physical science, but rather exists in the confines of the psychology of the ego.

The person or the ego is a noun that is attributed to a series of characteristics or features that remain present during the passing of time and can be defined or be classified within the physical or spiritual realm. However, personality is by its nature always changing, from the physical point of view as well as the psychological one; it is dynamic, evolutionary, and, like public data, it can be indefinable, strictly speaking.

The definition of a human being is considered appropriate from a public point of view, and we called this identity, when its verification has had a tendency to be based only on physical characteristics, such as photography and digital prints, until the modern-day invention of biochemical tests of DNA. There is a scientific possibility that this latter approach still may be rendered useless and insufficient to define identity should biological technology be able to introduce genetic modifications into adult individuals.

From a private point of view, individuality is a consequence of self-consciousness and it is bound with memory. The amnesiac seemingly loses his identity, at least to himself, even when, for example, for legal reasons—that is to say, reasons that are external or public in general—he can conserve it. Could it be possible that inversely the identity is conserved after physical death, when legally and objectively the opposite is defined?

We believe that memory and other many psychological attributes are inseparable from the physical organization of the cells that form our body. We can't imagine that neurons think individually, much less other cells of lesser "noble" tissues, such as those of the heart muscle. We consider these alien; they belong to the external world. We don't share our identity with the cells that give us form—neither with molecules like DNA, and much less with atoms and molecules that we exchange continually with the environment.

To social groups, we assign proper names, but we say that they are corporate entities—that is to say, they are ideal. Nevertheless, we evaluate ourselves in a "special" way; our nature is felt but not expressed. This inexpressible feeling has continuity; it is formed and it disappears, it is a characteristic of the world, it is at the same time collective and individual, it is eternal and temporary. It is transcendent and immanent.

We can imagine that some state of our individuality can evolve through successive lives, although this continuity doesn't correspond to public facts that could be established. The modality of the ego is infinitely changing, in a much more dynamic way, with less temporary permanency, than the physical "me." That mode of change could be constituted in a transcendent property that cannot be defined from a public focus, and that is manifested through successive individualities. On the other hand, what is the difference: to begin an apparent new identity through death, in an empty mind, or to change it every day starting from the day we are born?

Alternatively, we can reason that the desire for survival takes us to the illusion of the transmigration of souls, a doctrine that extends to the whole of nature in Hinduism. Consequently, we can consider these beliefs like a concession that is granted by the elites who cultivate a developed esoterism to religious organizations, in the name of popularizing the ethical impact they desire. However, in Buddhism, the absence of the ego is explicitly suggested, giving a fundamental character to this stand. It is worth observing that from the Platonic idea of the soul or spiritual substance that later became a part of the Christian doctrine, a contradiction exists in the transmigration of the identity and its negation when considering it fiction.

On the other hand, it is worth looking deeper into how the absence of the ego, considered illusory, conduces to spiritual continuity in nature; not from the substance point of view, nor from cause, but as a fundamental aspect that remains and evolves to the way of physical properties. When eliminating the ego, as noun, it becomes a modality, a form of that spiritual continuity that flows. Paradoxically, we can also consider the illusion of the ego as a yoke that nature imposes, in its eternal game that sustains the flow between opposites, to be or not to be. In that sense the "me" is eternal; it is an indispensable dynamism needed to sustain the game of the evolution of nature. Liberating us of its yoke brings us closer to God, the Tao; inversely, if we cling to it, we fulfill the purpose of life and come closer to the state of animality.

7-5 MORAL PERFECTIONISM AND SPIRITUAL TRANSCENDENCE

All the branches of Buddhism emphasize behavior, as much from the ethical pragmatic angle that seeks human well-being as an end in itself, and as a means for improvement and spiritual transcendence. The fundamental teaching is the suppression of desire that also implies the disappearance of the ego, individuality that motivates human suffering and misery. In Western understanding, including the two ideologies, spiritualist and materialistic, this ideology is considered contradictory, incoherent, because human well-being is based precisely on satisfying desires that differ according to philosophical conceptions, but that have the same ontological foundation, the omnipresent "me." In Western spiritualistic credos, individuality that transcends death is also decisive; and although it is certain that these credos propitiate altruistic behavior and compassion toward thy neighbor in the earthly stage, basically this behavior is looked upon as a transfer of pent-up desires to be rewarded in future stages.

Buddhist wisdom, axial as perennial philosophy, proposes the natural arc of human development, in its external branch, in accord with the law of existence that imposes on us the yoke of the ego and that also gives us the opportunity to halt and return in its involution, by means of illumination that dispels individualistic fantasy to its origin: Mother Nature, the Tao, Atman, God.

On the road of existence, the ego coexists with Tao's unit, with the vacuum and nonexistence, to the way of a wave that can be described by a Fourier series whose terms of infinite summation can take us, according to its preponderance, from hell, where the maximum frustration of desires of the ego prevail, to Nirvana, which by means of the illumination of transcendental meditation transports us to oceanic impersonality.

The materialistic understanding that prevails today in a dominant way in modern societies and that partly, at least, has its origin in Western science, which degrades in pseudo-scientific fashion, is not in itself better or worse than all the "isms" that have lashed humanity and that maintain our species close to animality, where the ego reigns tyrannically and could be the scientific superman or the God of eternal war.

The pseudo-scientific paradigm has convinced us that individual existence is unique; it will never be repeated in the coming of time, which is the new God that remains. Nobody doubts this unanswerable truth: We will never have the opportunity to live again; the ego disappears forever.

Only by means of meditation and cultivating detachment from desire, can we awaken to a new life, through the illumination that dispels the shadow of the "me" that is the means imposed to rise from the vacuum, which has the potentiality for existence.

Meditation on the false pretense of clinging to the concept of the ego allows us, momentarily or permanently, to achieve states that eventually are self-reinforced in a sense of ethical detachment and altruism, secured by universal knowledge. The wave-like tendency of nature can take us to very disparate feelings that are alternated in a random fashion, polarizing toward extremes of dissatisfaction or harmony in the peace of impersonal contemplation. Paradoxically, our choice is in the consciousness we want to forget. Saint Teresa of Avila tells us: "Such high hopes for life I have, that I die because I don't die . . ."

8

ETHICS AND POLITICS

8-1 IDEALISM AND PRAGMATISM

It is said that the Freemasons hold as sacred norm to speak neither of politics nor of religion, except during their secret meetings. This is very old wisdom, dating from the time of imposing Gothic cathedral constructions, silent symbols of the artistic wisdom of their architects, who captured their desires for freedom amid the fanaticism implied by the prevailing ideology.

Difficult terrain is this on which we traverse, as we descend from the regions of abstract thought inhabited by the gods, not to the cathedrals, but to the fortified castles and convents erected among the hills of the seven capital sins.

To what extent can the wisdom of the gods and their celestial court be transported to lands where desires, which nature inexorably has imposed on humanity, reign?

This query bears the great dilemma that the human species and life in general confront in their evolution, the trajectory of which is simultaneously individualistic and socialistic, selfish and altruistic. The biosphere remains evolving as a whole, by means of seemingly contradictory forces: the self who wants to be a unique God and its ephemeral destination, which eventually harmonize.

The dynamic coexistence of desire, an individual characteristic, with systemic cooperation, stochastic by nature, balances chance and necessity, the conscious and the subconscious, freedom and hierarchy; or alternatively, they veer off into another stadium toward a new game.

Let us note that these reflections, in the human world, have taken us to use a group of words, such as *conscious* and *subconscious*, *selfishness* and *altruism*, *desire* and *cooperation*, which are related to ethics, and, on the other hand, others such as *freedom* and *hierarchy*, *individualism* and *socialism*, which are in the realm of politics. We also use other words that might be called neutral, such as *chance* and *necessity*, *random* and *stochastic*, *systemic* and *sequential* or *linear*, which seem to be in the ideal kingdom of the wisdom of the gods, and of the universal or Platonic ideas, where we would not be subject to the reality that existence imposes on us.

In the world of ethics and politics we can remain within idealism and create doctrines and precepts to try to harmonize them, as if we lived in a celestial world, or confront them with realism, by means of the experience that each evolutionary stage imposes on us. To notice these alternatives, which we find today obvious, has not been precisely a characteristic of cultural development. On the contrary, traditionally, ethical or behavioral norms have been generally referred to as idealistic principles that take form in dogmatic paradigms that, once established, lose their ties to their origin of pragmatic character.

The history of religions and political ideas clearly illustrates this trend; ethical norms that refer naturally to the individual, such as Moses's ten commandments in the Bible, largely had a pragmatic origin, and were institutionalized by means of mythology of the personal God and man's original sin, with the main goal of exercising politics—that is to say, the establishment of coherent means for social organization that allow for its continuity. This process was carried out in a similar way in all civilizations, in a natural way, following in cultural-developmental paths close to those of the biological evolution, at least in its primitive stages. Mythology went along perfecting and rationalizing itself, passing to stages where philosophy and science had been developed in a less intuitive and more conscious environment.

The scientific induction process is outlined in a cultural development, as far as facts and social results are harmonized with the proposed ideology. An ideological position eventually configured will induce political stability, if it adapts to internal and external dynamic factors that are exerted on a society or a civilization in its evolution. Nevertheless, there is a fundamental difference between the development of human knowledge and social and political development. In the acquisition of knowledge, preferably if it refers to the physical world, its evolution strictly refers to facts that are easily subject to experimentation and confrontation; this is not the case with social phenomena, which develops during long periods of time during which changes are not simply noticed. Also, in politics it is sought to influence social facts by means of the application of an ethical ideology; in logic and science the facts are unchangeable. For these reasons, fundamentally, humanity's ethical and political evolution has always lagged behind philosophy and science. This situation has been a historical constant to date, as we can verify better than never, how they diverge exponentially.

Ethics and politics are not in the same world as logic and science, but in the world of desires and feelings. Only historical reflections can show us, in a very limited way, how to use the acquired social experience to project into the future, always keeping in mind that the key is not in determining goals, which are in the fundamental ethical definitions, but in the means that are used to reach them. Methods used in natural sciences cannot strictly be transposed to politics; ethical ideology has a pragmatic character that has to adapt to changing societal conditions and strictly cannot depend on supposed ideological principles, to the way of mathematical axioms and science, nor from alleged mythology. A look into some historical developments facilitates ethical and political understanding and supplements the analysis of ethical ideologies simplistically intended to address systemic problems of human behavior.

8-2 SELF-ORGANIZATION OF THE EVOLVING POLITICAL HISTORY

In Western civilization, there have occurred two major cultural transformations: the blossoming of Greek philosophy, starting from the sixth century before the Christian era, and the European Renaissance. In both stages we can see important transformations of the cultural paradigm, which have an individual ethical texture, and an organizational or political behavior, which gives continuity and temporary permanency to society. Only by means of a complex mutual relationship among these systems—individual ideology and social organization—is it possible to stabilize certain social patterns that define those civilizations. Scientific mentality and political democracy were born in ancient Greece; the grade of social development they reached was limited by the ethical and economic levels they achieved. Democracy was confined to a geographical and social space defined by the city-state and to a privileged class, the Greek citizens. The struggles among rich and poor and the external threats were insurmountable; physical means that would have allowed them to go on further didn't exist.

The establishment of the Roman Empire made the expansion of civilization possible geographically, at the cost of sacrificing democracy, even at the reduced level of the Greeks, who considered a need for slavery of foreigners and prisoners of war.

The empire contributed notably to cultural diffusion and to the improvement of the law and technology, sacrificing cultural and scientific innovation. The ethical advances, as the ideals of equality lost ground, gave place to Christian religious thought that claimed, by means of its theological myths, a new ethical order that projected the equality of men before a unique and just God. The ethical-political ideals moved beyond, to the eternal life, propitiating a theocracy that forcibly shared power and cultural heritage of the empire with the barbaric warriors that caused its breakup. This duality characterized political power with religious preponderance of the church that possessed the traditional unifying cultures, confronting the warring and geographically disperse power that slowly was consolidated feudally during ten centuries of the so-called Middle Ages. The strengthening of trade in Italy and the discovery of America contributed to the growth of a new bourgeois class that made possible economically the cultural development of the fifteenth and sixteenth centuries.

From the European Renaissance, a new spirit, based on what today we call the scientific method, started taking shape, as it tore down philosophical ideologies and religious beliefs that were assumed to be unchangeable covering all cultural aspects of Christian civilization. The moderation of strict ethical standards resulted, in connection with the arts, in a renovating and creative expression manifested in all of the Renaissance's literary, musical, and plastic dimensions. Science took an absolutely new and creative direction, when it heretically ignored not only the prevailing Christian doctrine based on the texts of Jewish origin, but also dared to question the Greek philosophical authorities, deified from the days of Aristotle.

For the Greeks, ideology governed the world: It was said, if the facts don't agree with philosophy, that just indicated their imperfection; the ideal kingdom of the gods, Platonic ideas and numeric Pythagorism, should prevail. On the other hand, at the end of the Middle Ages, the Greek logic, dusted off by Thomas of Aquino, was placed at the disposal of Christian dogmas, making it possible for the scholastic to give Hebrew mythology a robe of rationalism, while paradoxically contributing to its questioning.

During the Renaissance, a new ideological synthesis arose that led to a substantial change in the methodology used for acquiring knowledge: The

Greek mathematical-logic united with the so-called principle of objectivity, which subjects ideology to a confrontation with the facts of nature. This philosophical attitude marked a new direction for the whole conception of Western civilization, making possible the spectacular development of physical science. In parallel, albeit at a slower pace, religious dogmatism was losing the absolute dominion it exerted at all ideological levels, giving place, with the influence of the new science of economy, to substantial political transformations.

All of this philosophical and scientific development would not have been possible if the political conditions of European society had not evolved favorably, diminishing the influence that the prevailing ideological dogmatism had over secular power. Religious struggles with the Roman Christian Church, inspired and directed mainly by English and Germanic kings, didn't really transform Christian dogmatism, but rather reduced its political power and made possible the flourishing of scientific heresies that were timidly being reinforced. The cultural movement of the Renaissance that arose initially in Italy moved toward the north, where the victorious religious Protestants exercised less political influence than the traditional Christian church.

These historical events show the systemic mechanism that clearly characterizes social transformations: The relationships between ethical principles and political facts are mutual, as with every evolutionary process. The patterns that typify their development will occur, with more or less speed, in a way that cannot be described sequentially, as cause and effect is in mechanics. Far from classic determinism is the environment of the stochastic nature of these transformations, which are not wholly random nor are they fatally defined. It is necessary to give up our own pretense of simplistic ways of thinking that establish eternal truths describing social events by means of the "immutable laws of nature."

Models from the mathematical theories of chaos simulate in an appropriate fashion the succession of structures or dynamic patterns, which are described by means of "valleys" that hold their trajectories, which are subjected to stochastic transformations of mixed nature—that is to say, deterministic and random. In social models, we idealize ethical relationships that are established by means of individual psychological behavioral mechanisms, and relate them to the social whole, that in a web-like fashion, by means of communication among its components, is governed by political norms that lead to an evolutionary social behavior.

In political history a constant can be visualized consisting of a duality of factors of power, which allow social cohesion. The novelist Aldous Huxley, in his political essays, characterizes them in a singular form: "Church and State, Greed and Hate, Two baboon-persons in one Supreme Gorilla." It is curious to compare ourselves with our near cousins of the animal kingdom, by attributing to them our biggest defects. In a benign way, we can interpret this ethical similarity, this reference to animalism, as an expression of the Darwinian law of survival of the fittest that contrasts with the supposed altruism we kindly attribute to ourselves. Let us not forget that cooperation also exists in the animal kingdom, as is evidenced in biological studies of ant societies, or in ecosystems of greater complexity. Nevertheless, let us return to a space within our boundaries, considering that in the animal kingdom everything happens as an evolutionary chance event, which in turn derives to instinctive behavior or automatism, contrasting with our conscious human nature that allows us to choose, to make decisions—in other words, free will. We believe that this exclusive characteristic of human beings is the ideological basis of morality or ethical behavior. It is worth asking if consciousness, in an ethical sense, belongs solely to man as an individual or may also extend to human societies as a whole.

Intuitively, common sense indicates that the concept of freedom is considered an individual attribute that can hardly be extended to the collective. However, we establish a clear relationship between them when we refer to ethical norms that can also be considered as generalizations. In this way we arrive at the political environment, where laws govern our destinies, and we find the origin of the institutions of power to which Huxley refers. Are they fruits from capital sins, such as greed and hatred? Western civilization, following the schism of Christianity called the Reformation, found a new path to social organization, which has been identified as the separation of church and state, called "laitism" or "secularism." The great thinkers of this modernism have directed the philosophical current toward politics, trying to give it a rationalistic conception. How fortunate has this direction been which has characterized the last five hundred years of Western history and has extended to the whole orb?

Let us note that this separation of church and state doesn't guarantee in itself the modern democratic idea whose evolution and improvement depends fundamentally on a plurality and independence of the centers of power, sustained in an anti-dogmatic attitude, in all cultural dimensions. Philosophers such as Thomas Hobbes in the seventeenth century proposed absolutism of the almighty state, eliminating the influence of religious ideology especially from the mythological point of view. This conception was justified as a necessity to maintain order and material progress, from a pragmatic point of view only. It is worth mentioning that this political doctrine basically still subsists, four centuries later, especially in countries of the "Third World." The most outstanding representative is modern-day China, which has eliminated its traditional ethical values, such as Buddhism, to build a powerful, industrialized society, in its materialistic emulation of the West. Its empirical nature is characteristic of this philosophical-political understanding, which is close to the pseudo-scientific philosophy transferred from the world of Newtonian mechanics to sociology.

This strengthening of the laic state was not in itself a political novelty; its creative aspect is related to its ideological independence, to the way of the new science that rejected the prevailing religious dogmatism of the Middle Ages. This attitude, reinforced today, is not exempt from returning to a new pseudo-scientific dogmatism, which threatens to submerge humanity into a new Dark Age. The religious struggles of the Christian Reformation weakened ecclesiastical power in favor of secular power, which looked for a new ideology to sustain the new organization of the state. The adopted pragmatism has been a new ethical attitude, but it is not a theory or belief that can be established as an element of social cohesion. The search for this ideological foundation has characterized Western political-philosophical thought to present times.

The evolution of that modern paradigm has been accelerated extraordinarily by the emergence of industrialism, which, in introducing a revolutionary new technology, economically has radically transformed the physical and social environment. Ideologies have emerged in rapid flooding succession in a historical testing ground that doesn't allow time for their evaluation, taking humanity to the edge of self-destruction in the twentieth century, characterized by devastating wars, the fruit of ideologies based on alleged "historical laws" that, to the way of theories of natural sciences, were considered relentless. The most outstanding of those philosophical ideologies, in terms of the political impact they have had, were elaborated by G. W. F. Hegel and Karl Marx in the eighteenth and nineteenth centuries. These ideologies gave place to National German Socialism and Stalinist Communism, which were characterized by the magnification of the state as a nationalistic and messianic force directed toward world dominance.

8-3 THE CONTEMPORARY PSEUDO-SCIENTIFIC PARADIGM

Among the simplifying myths that have been taking hold of popular political mentality, stands out the classification of political ideas by means of the one-dimensional geometric simile. Through the media, political forces have attempted to indoctrinate us, apparently with success, using geometric vocabulary—left, right, and center—to designate political theories and consequently the political movements, or parties. At first sight this simplistic classification may seem innocuous; nevertheless, it represents a polarized approach to ideological evaluation that is frequently used to impede dialogue that should prevail in the creative exchange, especially in matters as controversial as politics and ethics.

It would be convenient if this Cartesian simplification, which seems to have caught on so well in the political classes, was taken to multidimensional diagrams or at least to a plane. Certainly, such representations would not be advisable to their employment in the public square, but at least the "political actors" could use them in gatherings of some sophistication, where some rigorous analysis was demanded. By way of example, we can consider two-dimensional representations, choosing appropriate political variables according to the application that is sought to give them. In the divisions of public power in modern societies, we consider as fundamental the branches called executive, legislative, and judicial, which on the whole can designate "political power" corresponding to the state, and other entities related to production, such as industry and trade unions, which we label as "economic powers." Within those sectors exists an infinity of other centers and factors of power in a pluralistic society, which could be considered from a wide perspective, among which stands out the so-called third power or the media, which for purposes of our example, we can assume that the ingenuity of the analysts has included equally among the two powers, political (P) and economic (E).

When applying the Cartesian system of representation we should choose the variables, and in our example we would pass from the linear limitation to a plane, where we can try to locate political societies and their trends. In a linear representation, all variables are superimposed and the only thing that is expressed is their intensity in one direction or its opposite. What is meant, when the ideas of a speaker, a political actor, or a party are classified as extreme right or left, or of the center? For discussion purposes, we can define the distribution of political power (P) and economic power (E) among the members of a society, by means of Cartesian power variables designated as dP and dE. Certainly these definitions should be quantified strictly, but for our purposes we will consider them in a semi-quantitative base of representations within a plane, as in Figures 31 (a), (b), and (c).

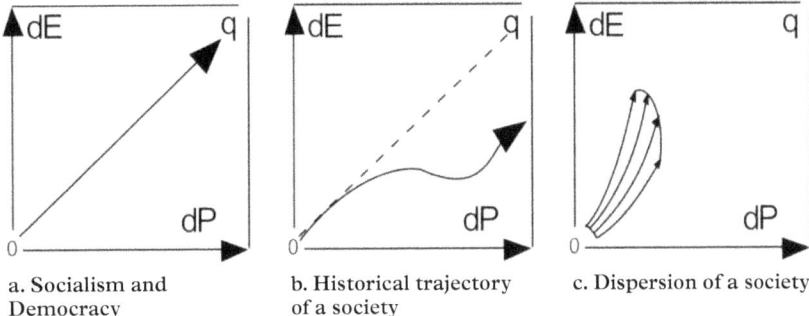

a. Socialism and Democracy

b. Historical trajectory of a society

c. Dispersion of a society

Figure 31: POLITICS IN TWO DIMENSIONS

In these figures, the origin (0) corresponds to an extreme concentration of political and economic power in a ruling elite that we can call an oligarchy or government by few. The point (q) of the opposed vertex refers to the maximum distribution of the wealth, or socialism, simultaneous with a government of shared political power or a democracy. Points on the axes represent societies, where a distribution of power P or of power E could exist exclusively; in contrast, the diagonal represents balanced societies. Marginally, seeking a better understanding of this geometric-political representation, as in Figure 31(b), we are shown a line or a trajectory that, parametrically with time, represents the historical evolution of a social conglomerate, and in Figure 31(c), the dispersion phenomenon that might occur in any given society, within a political field characterized by some determined lines of flow. The possibilities for these geometric representations of social evolution are promissory, even though their reliability should not be considered in a mechanistic context to the style of natural sciences, but simply as a complementary element to this difficult task.

It is not possible to represent two or more independent variables in a single dimension; we should in principle use the Cartesian analysis. The only possibility would be to define in our case a variable $\pi = f(dP,dE)$, that somehow simultaneously implemented the two powers, economic and political, and would characterize the cultural-political development of a society. In fact, we know that this relationship can be expressed, which describes history or social trajectory, and is expected to contribute successfully in dispelling totally the political classification we have called one-dimensional. This classification has an ancestral, distant origin, imprinted in the language itself: The West traditionally has associated the right with correct ethical behavior and the left with the opposite. In the Christian credo, it is said that Jesus after the resurrection was seated to the "right" of the Father. In common language, *sinister* (an old English word meaning "on the left side") is synonymous with *perverse* or *nefarious*, and opposite of *right* ("dexter"), or *skilled* and *sagacious*. In political protocol—not in the game of chess—the queen is placed to the king's right; in modern days this convention is being forgotten, even though it seems to be resurging "dialectically" when assigning to the "left" a political paradise.

It is worth wondering if this modern, linear special designation is a consequence of the influence of Hegel and Marx's philosophical theories that interpret history dialectically . . .

The dissident French political pundit Jean F. Revel[1] presents in *The Totalitarian Temptation* and, more recently, in *The Great Masquerade* an interesting analysis of these aspects of political power and its tendencies after the Second World War. In the first book, he refers to the political illusion that arose in the postwar period as a consequence of the polarization induced by the "Cold War," which favored the concentration of power. On the other hand, the latter book refers to the glorification of the "left" that the author defines as a masquerade, a political deceit orchestrated by a worldwide journalistic core that defines *a priori* "political advancement" because of the ideals of equality it claims to represent, without keeping objectively in mind the results of its actions. It seems as though the eternal treadmill of political history is returning to the Romantic idealism of Rousseau that hides, consciously or unconsciously, the current destruction of the fleeting democracy that in fact disappears when its supposed defenders move away from the principle of objectivity.

If there is a philosophical lesson to infer from the current crisis in the ethical and political environment, we should consider the failure of ideologies that have tried to substitute myths that traditionally sustain a certain political cohesion, by means of religious doctrines. It is not intended by any means to return completely to the times we have already overcome, which were based on superstition and ignorance. In the search for new horizons, an attitude of pluralistic ideological coexistence is fundamental, one that recognizes diversity as an inexhaustible resource for understanding a complex, dynamic world that is not governed by inflexible norms, but by subtle rules, the organization of which we should bring to light by means of study, without hop-

---

[1]Jean F. Revel, *La Tentación Totalitaria* (Plaza & Janés Ed. 1976).

———, *La Gran Mascarada, Ensayo Sobre la Supervivencia de la Utopia Socialista* (Editorial Taurus 2000).

ing to find them in a definitive form, but, on the contrary, recognizing their evolutionary nature. Social transformations are not created by destructive political revolutions, but by interactions that modulate great transformations that are imposed evolutionarily. Huxley, employing biological similes, states wisely: "the aims are simian election, the means are only human choice"—the means determine the ends. The French Revolution is considered the mother of current democracy; an objective analysis of history shows us that France arrived at political stability through a much longer and more tortuous road, through emperors and successive upheavals, while the Anglo-Saxon nations acquired higher maturity, step by step by means of an evolutionary process. It is hoped that man's rights, "*Liberté, Egalité, Fratenité,*" of the Encyclopedists, are also made a reality in the economic context, by means of a realistic evolution of democratic socialism, keeping in mind the revolutionary disaster of soviet communism.

The modern democratic model, which characterizes societies of the dominant countries in the intellectual and economic context of the twenty-first century, represents valuable ethical and political achievements of mankind, which by its own nature is called to exercise an important leadership role in the social and economic evolution of the whole planet. Today's fundamental challenge is the globalization of the political culture already reached by means of understanding the evolutionary nature that makes possible a process of this magnitude.

In today's world, economic globalization is spoken of as the key factor for an opening directed toward humanity's well-being. Certainly, the importance of the development of the necessary physical means to acquire a minimum base that propitiates the individual and collective "happiness of mankind" cannot be ignored. Still, admitting the high priority of this supposition, it is necessary to evaluate it qualitatively, identifying the physical means that are an absolute priority for human well-being. In current historical circumstances, as growing world trade is being impelled, especially among the developed block of countries and the rest of the world, trade is started from an asymmetrical bargaining position like the geometric figure of a funnel, in which not even equal bases are established for the parts, the minimum required conditions to obtain good results. On the contrary, the exchange rules seek to favor the strongest, and are established this way, even turning back to the status quo. In addition, the base of the argument contradicts itself in principle, which supposedly searches to redeem the large human conglomerates that remain below the minimum threshold of physical survival.

World trade is currently oriented to sell agricultural surpluses that rich countries offer at subsidized prices by means of exaggerated profits obtained from the sale of high-technology goods. As if the spoliation of less developed countries were small, as they receive industrial trinkets that flood the world markets, protected by the commercial opening that in this form contributes nothing to their well-being. Let's not be surprised by the deplorable results: unemployment and therefore misery pervades these countries, which are even denied the right to produce their daily sustenance, alleging inalterable economic principles, called laws of the market, according to which they should literally die from hunger while watching their fields idle.

It is said that free trade treaties favor poor countries because they allow them to acquire, at cheaper prices, their basic necessities such as grains, hence devoting their efforts to exporting those products that, because of local climatic conditions and the use of cheap labor, can be produced competitively. This entrapping narrative repeats itself clearly in economic world history through examples such as the world market for coffee, rubber, and so many other products. What is going to happen to the trade of flowers and exotic fruits, as their production spreads? An economic collapse will occur in countries that don't even produce their basic food requirements, because they have not acquired the necessary technology to mass-produce them at acceptable prices. Could it be possible, alternatively, that instead of exporting agricultural goods of sumptuary consumption, the developing countries receive technology and capital by means of joint ventures that would allow them to compete industrially on a global level? There is no doubt that this path offers better possibilities to promote economic well-being, along with the development of international tourism.

Paradoxically, among the determined opponents of so-called globalization, important ideologies of opinion are present, especially in Europe and in some North American sectors that, by hiding their national and group interests, seek to deceive developing countries with their alleged altruistic defense of social vindication. To maintain or increase the commercial and labor privileges they traditionally have enjoyed, it is not in their best interests to lower subsidies or reduce prices for high-technology goods that may be subject to competition from new developments that eventually arise in geographical areas they can't control.

The current economic opening of world markets fails at its foundations—not by its physical, industrial, or capital principles, but by its ethical principles, which are based on the laws of the jungle, on the Darwinian survival of the fittest. The illusion of unlimited economic growth, which is considered as another "scientific" postulate, is sought to be maintained at all costs in developed countries, who contemplate dumbfounded the new slaves that try by all possible means within their reach to pass over the threshold that separates them from the "paradises" where wealth is accumulated.

Bringing to our political analysis this economic aspect of modern times allows us to glimpse the complexity of the factors involved in the evolution of human society. Again we can see that the formulation of simplistic theories that transfer outdated mechanistic methods from natural science to social phenomena takes us away from reality. We must recognize that it is not possible in the world of complexity to enunciate theoretical generalizations of any kind, which allow us in a messianic fashion to elaborate social plans, economic projects, or models of physical and cultural development that, inflexibly established *a priori*, drive us to utopian paradises.

History demonstrates to us clearly the failure of ideologies that seek by means of simple mechanistic arguments to discover utopian political models that redeem humanity. Greek democracy didn't reach farther than the geographical limits of the city-state or the social environment of the elite who excluded the foreigner who provided the enslaved labor class, justifiable at its roots by the need for territorial and personal defense. The grade of technological development of those times and the cultural abyss that separated Greek society from the "barbarians" made impossible the continuity and extension of that political model, which has now resuscitated twenty centuries later as technical means have made it feasible for the first time in human history to expand economic development for most of the population. Again, at the present time, we live in a historical situation similar to the one that caused the transition from the Greek culture to the establishment of the Roman Empire.

The so-called developed countries have achieved, after partially overcoming the social mechanistic models arisen in the West during the Industrial Revolution that led to the great wars of the twentieth century, an economic well-being that is shared by a great majority of their populations, in contrast with the rest of the world. In these countries, the distribution of wealth and the collective experience acquired during centuries of development have made possible the flourishing of Greek political ideas of democracy, evolving to a level never reached before. What are the factors that presumably have prevented most of the world's population from achieving the globalization of that successful pattern reached by the developed countries?

8-4 LOOKING FORWARD

Based on historical experience we can configure some ideas in this respect that bring to light the difficulties we should overcome. Above all, the inexperience of the underdeveloped world should be avoided, so as not to drive us to conflicts caused by political ideologies that have already been resolved by countries that suffered them in all their consequences. Great efforts will be necessary in the educational field to reduce the persistence of a primitive ethical mentality that still subsists in a generalized manner throughout humanity, which has its deep roots in human nature itself, which refuses to recognize the limitations of the ego that moves from the psychological to the political environment, giving justification to exclusionary nationalism. Selfishness, usury, greed, and all the capital sins that at least in theory are not admitted at an individual level, have subsisted in relationships among democratic states.

In democracy's current social outline has been deliberately excluded the individual ethical element that traditionally had a mythological base, socially imposed by organized religions. Actually, the prevailing ethical paradigm has arisen from the resulting void, taking shape in the form of extreme pragmatism, derived from the mentality created by the success of scientific technology in the domain of nature, where physical well-being prevails as the sole ideal, without recognizing any other value worthy of consideration. The individual and collective economic objective is manifested in a wild race after the acquisition of objects and wealth that accumulate indefinitely, to enable individual and collective power.

This modern paradigm takes man backwards to animalism, depriving him of his most valuable gift, his capacity to dream of transcending the ordinary life of physical subsistence by means of intellectual and moral development—his spirituality.

Let's not surprise ourselves with the disappointment of youth, which characterizes the most developed societies, where the lack of transcendent spiritual values creates the desire for fantastic experiences provided by hallucinatory drugs consumed in massive quantities.

Neither should we ignore that in underdeveloped countries, the frustration that produces the contrast between opulence and extreme poverty, as well as the promotion of unreachable economic luxuries, contribute to the endless wars that impede in permanent form the intended development proposed by the political and economic elites. We cannot expect otherwise from the prevailing materialistic paradigm: Everything is justifiable to reach individual and collective economic power; there is no other goal.

The action of a reduced intellectual environment seems to be extending to social nuclei, which eventually may acquire the so-called "critical mass" that determines an inflection of the current ethical-political trends that presage social catastrophes at a global scale. A new synthesis that reconciles religious and scientific mentalities, as opposed to one another as they may seem, has the potential of reorienting individual and collective ethical mentality toward recognition of human dualism. These understandings subsist precariously at the present time, without there being a sincere effort to produce a valuable approach. Sincerity is in fact the basis of ethics; it is not enough but it is necessary. Ethical principles are not only those dictated to us by biology, in its great diversity of ecological organization, but they should also travel the cultural evolutionary stage that differentiates the human species from the rest of nature, without forgetting our distant origins.

If we aspire at all to locate ourselves within angelical space, above good and evil, we should allow the cosmic oscillation that brings us closer to those divine strata, and relates us with the purely biological level, where the Darwinists, first cousins of the neoliberals, follow the principles or laws of the jungle. The struggles between ethical-religious and scientific-utilitarian attitudes have continued in the West until today, in a latent and irreconcilable fashion, leading to a social dichotomy between formal religiosity of appearances, and a pragmatism that is inconsistent with the ethical bases that it ideally sustains.

Pacifism as well as other ideologies or generalizations of human rationality, such as political democracy and economic communism, are feasible until they reach the threshold of the ego, which has a primitive ancestor we designate as animal. This ego constitutes an inherent force of nature, without which life would have not arisen. When we judge our neighbor, we cannot use idealisms, regardless of their personal jurisdiction which is inviolable as his own. This moral principle or behavior is not a law of nature; it is simply a consequence of the belief that there are other minds of their own, and probably we are alike. The human duality between the instinct for survival and the superconsciousness that projects us toward idealization cannot be dissolved without ignoring our reality: We are neither angels nor beasts; peace and war, despotism and liberalism, capitalism and socialism—these are faces of the same phenomenon that is the life. Only by means of a realistic, objective attitude that keeps in mind this duality would we be able to evolve mentally toward a stable social balance.

A convergence among those seemingly contradictory positions is possible and convenient; it becomes necessary for the current tendency of the utilitarian and pragmatic education to recognize the importance of spiritual values in the formation of the human mind, and, on the other hand, that religious organizations descend from their dogmatic pedestal, toward a mystic understanding, lacking of historical myths, recognizing a transcendent position.

The laic understanding of the separation of church and state has suppressed the study of religion in public education and has largely substituted it with nationalism. The results are catastrophic: The territorial animal instinct is exasperated and the justification for war is induced. In addition, an empty spiritual mentality at an individual level is propitiated, inclined toward practical superstitions that were believed to have already been overcome. Certainly, it is not meant that we must return to the educational confessionalism; on the contrary comparative study of religious ideas, deepening in their historical and ethical importance, will encourage a new way of thinking to achieve understanding and human solidarity.

Ethics should not be considered as a revealed truth; dogmatic and inflexible, its origin is evolutionary. It should conserve a certain degree of utilitarianism that doesn't suffocate our altruistic ideals or destroy individual freedoms and intellectual motivations that give meaning to our existence. Its principles should reflect biology in its great diversity of ecological organization, and fundamentally recognize the evolutionary stage that differentiates the human species from the rest of nature in its multiple spiritual facets. The ignorance of that duality of man's material and spiritual characteristics, as an individual and as a species, has greatly slowed humanity's ethical progress, in comparison to the great developments of science and technology, the main fruit of its rationality.

Ethics and politics veer away from logic, because they are tied to human desires that are voluble individually and collectively. It doesn't seem that collectively human societies, as such—especially at a global level, as it is proposed today—have the capacity that we attribute to ourselves individually to define our own destinies. Nevertheless, we may be optimistic, if we recognize that human history has experienced big transformations, induced by powerful individual wills, to the way of the "butterfly effect" that is expressed in the theory of chaos: The flight movement of a butterfly is amplified within a great storm that extends thousands of miles away.

Searching for an epilogue to this existential drama of human nature, Emilio Pradilla-González's poetic dimension comes into my imagination, from his poem called "Gypsy":

Freedom I conceive but no longer look for it.
My jail travels with me wherever I may go,
as with him always goes on the barren beach,
the pearly shell, harder, of the mollusk.[1]

---

[1]Emilio Pradilla-González, *Selección de Poesías, Prosa Lírica y Ensayos* (Bucaramanga, Colombia: Sic Editorial, 2009).

BIBLIOGRAPHY

Anderson, Jay M. *Mathematics for Quantum Chemistry.* W.A. Benjamin, 1966.

Andrade, Eugenio. *Los Demonios de Darwin.* Bogotá: Universidad Nacional de Colombia, 2003.

Asimov, Isaac. *A Short History of Chemistry.* New York: Doubleday, 1965.

Assmann, Hugo. *Placer y ternura en la educación: Hacia una sociedad aprendiente.* Madrid: Narcea, S. A. De Ediciones, 2002.

Bateson, Gregory. *Steps to an Ecology of Mind.* Chicago: University of Chicago Press, 2000.

Bechtel, William. *Philosophy of Mind: An Overview for Cognitive Science.* Hillsdale, NY: Lawrence Erlbaum Associates, 1988.

Beiser, Arthur. *Concepts of Modern Physics.* New York: McGraw-Hill, 1963.

Bernstein, Jeremy. *The Tenth Dimension: An Informal History of High-Energy Physics.* New York: McGraw-Hill, 1989.

Bohm, David, and F. David Peat. *Science, Order, and Creativity.* New York: Bantam Books, 1987.

Capra, Fritjof. *The Tao of Physics.* New York: Bantam Books, 1983.

———. *The Web of Life: A New Understanding of Living Systems.* New York: Doubleday, 1996.

Chopra, Deepak. *Ageless Body, Timeless Mind.* London: Rider, 2003.

Coulson, C. A. *Waves: A Mathematical Account of the Common Types of Wave Motion.* London: Oliver and Boyd, 1961.

Cruz, Manuel de la. *Filosofía contemporánea.* Taurus Ed., 2002.

Dalai Lama. *The Universe in a Single Atom: The Convergence of Science and Spirituality.* New York: Morgan Road Books, 2005.

Deutsch, David. *The Fabric of Reality: The Science of Parallel Universes and Its Implications.* New York: Penguin Books, 1997.

Echeverría, J. *Introducción a la metodología de la ciencia.* Barcelona: Ed. Barcanova, 1989.

Ellis, George F. R., and Ruth M. Williams. *Flat and Curved Space-Times.* Oxford: Clarendon Press, 1990.

Eyring, Henry, John Walter, and George E. Kimball. *Quantum Chemistry.* New York: John Wiley & Sons, 1965.

Feynman, Richard P. *QED: The Strange Theory of Light and Matter.* Princeton, NJ: Princeton University Press, 1985.

Feynman, Richard P. et. al. *Lectures on Physics.* Boston: Addison Wesley, Longman, 1970.

Ferrater Morta, J., and H. Leblanc. *Lógica matemática.* Mexico: Fondo de Cultura Economica, 1994.

Ferry, L., and J. D. Vincent. *¿Qué es el hombre?* Madrid: Editorial Taurus, 2001.

García Belmar, Antonio, and José Ramón Bertomeu Sanchez. *Nombrar la materia: Una introducción histórica a la terminología química.* Barcelona: Ediciones del Serbal, 1999.

Gibran, Kahlil. *La Procesión.* Barcelona: Ramos Majos, 1982.

Gleick, James. *Chaos: Making a New Science.* New York: Penguin, 1988.

Goswami, Amit, Richard E. Reed, and Maggie Goswami. *The Self-Aware Universe.* New York: Penguin Putnam, 1995.

Gray, Jeremy. *Ideas of Space: Euclidean, Non-Euclidean, and Relativistic.* New York: Oxford University Press, 1989.

Hawking, Stephen W. *The Theory of Everything: The Origin and Fate of the Universe.* New Millennium Press, 1996.

———. *The Universe in a Nutshell.* New York: Bantam Books, 2001.

Hawking, Stephen W. et al.: *The Future of Spacetime.* New York: W.W. Norton, 2002.

Hillmer, Rachel, and Paul Kwiat. "A Do-It-Yourself Quantum Eraser," *Scientific American* 296, no. 90 (May 2007).

Humphrey, Nicholas. *A History of the Mind.* New York: Simon & Schuster, 1992.

Huxley, Aldous. *The Perennial Philosophy.* New York: Harper, 2004.

Ikeda, Daisaku. *Life: An Enigma, a Precious Jewel.* Kodansha International, 1982.

Jung, Carl G. *The Archetypes and the Collective Unconscious. (Las relaciones entre el Yo y el Inconsciente.)* Ediciones Paidós, 1997.

———. *Synchronicity: An Acausal Connecting Principle.* Bollingen, 1973.

Kauffman, Stuart. "Antichaos and Adaptation." *Scientific American.* (August 1991).

———. *At Home in the Universe: The Search for the Laws of Self-Organization and Complexity.* New York: Oxford University Press, 1995.

———. *Investigations for General Biology.* New York: Oxford University Press, 2000.

Kellert, Stephen H. *In the Wake of Chaos.* Chicago: University of Chicago Press, 1993.

Kenny, Anthony. *The Metaphysics of Mind.* New York: Oxford University Press, 1989.

Kilmister, C. W. *Russell.* Hampshire, UK: Palgrave Macmillan, 1992.

Kwiat, Paul, and Berthold-Georg Englert. "Quantum Erasing the Nature of Reality, or, Perhaps the Reality of Nature?" *Scientific American* (May 2007): http://www.scientificamerican.com/article.cfm?id=a-do-it-yourself-quantum-eraser.

Lao-Tse. *Tao Te King.* Translated by Roberto Pla. Mexico City: Editorial Diana, 1980.

Lawden, D. F. *An Introduction to Tensor Calculus and Relativity.* Redwood Press, 1967.

Lawrie. Ian D. *A Unified Grand Tour of Theoretical Physics.* Bristol, UK: Adam Hilger, 1989.

Lewontin, Richard. *It Ain't Necessarily So: The Dream of the Human Genome and Other Illusions.* New York: New York Review of Books, 2000.

Lindley, David. *The End of Physics: The Myth of a Unified Theory.* New York: Basic Books, 1993.

Llinás, Rodolfo. *I of the Vortex: From Neurons to Self.* Cambridge, MA: MIT Press, 2002.

Loeb, L. B. *Fundamentals of Electricity & Magnetism.* New York: John Wiley & Sons, 1947.

Marinoff, Lou. *The Middle Way: Finding Happiness in a World of Extremes.* New York: Sterling, 2007.

———. *Plato, Not Prozac: Applying Eternal Wisdom to Everyday Problems.* New York: HarperCollins Publishers, 1999.

Margulis, Lynn, and Dorion Sagan. *Microcosmos: Four Billion Years of Microbial Evolution.* Berkeley, CA: University of California Press, 1997.

Maturana, Humberto. *La realidad: ¿objetiva o construida?* Anthropos Editorial, 1997.

Menaker, Esther, and William Menaker. *Ego in Evolution.* New York: Grove Press, 1965.

Monod, Jacques. *Chance and Necessity.* New York: Vintage Books, 1972.

Murdoch, D. R. *Neils Bohr's Philosophy of Physics.* Cambridge: Cambridge University Press, 1987.

Pauling Linus, and Edgar B. Wilson. *Introduction to Quantum Mechanics.* New York: McGraw-Hill, 1935.

Penrose, Roger. *The Emperor's New Mind.* New York: Oxford University Press, 1989.

———. *The Road to Reality.* New York: Random House, 2004.

———. *Shadows of the Mind.* New York: Oxford University Press, 1994.

Piaget, Jean. *El desarrollo de la noción del tiempo en el niño.* Mexico: Fondo Cult. Ec., 1992.

Pilar, Frank L. *Elementary Quantum Chemistry.* New York: McGraw-Hill, 1968.

Popper, Karl, and John C. Eccles. *The Self and Its Brain.* Routledge, 1984.

Pradilla-González, E. *Selección de poesías, prosa lírica y ensayos.* Bucaramanga, Colombia: Sic Editorial, 2009.

Pradilla-Sorzano, Jaime. "Spectroscopic Studies of Cu(II) Complexes." Ph.D. Thesis, Case Western University, 1972.

———. "Infra-red Spectra and Normal Coordinate of Pt(II) Complexes." M.Sc. Thesis, Case Institute of Technology, 1966.

Pradilla-Sorzano, Jaime, and John P. Fackler, Jr. "Base Adducts of ß-Ketoenolates. VII. Electron Paramagnetic Resonance Studies of Some Fluxional 1,1,1,5,5,5-Hexafluoro-2,4-pentanedionatocopper(II) Complexes," *Inorganic Chemistry* 13, no.1 (1974): 38–44.

———. "Base Adducts of ß-Ketoenolates. VI. Single-Crystal Electron Paramagnetic Resonance and Optical Studies of Copper (II)- Doped cisBis(hexafluoroacetyla cetonato)bis (pyridine) zinc (II), Cu-Zn (F_6 acac)$_2$(py)$_2$," *Inorganic Chemistry* 12, no. 5 (1973): 1182–1189.

———. "BaseAdducts of ß-Ketoenolates. V. Crystal and Molecular Structures of cis-Bis(1,1,1,6,6,6-hexafluoro-2,4-pentanedionato)bis (pyridine)zinc(II) and – copper(II)," *Inorganic Chemistry* 12, no. 5 (1973):1174–1182.

———. "Far Infrared Spectra of Olefin-Pt(II) Complexes. Normal Coordinates Analysis of Zeise's Salt," *Journal of Molecular Spectroscopy* 22, no. 1 (1967): 80–98.

Pradilla- Sorzano, Jaime, H. W. Chen, F. W. Koknat, and John P. Fackler, Jr. "Structure and Electron Paramagnetic Resonance Spectrum of the Product of the Reaction of Aqueous Pyridine with Copper (II) Hexafluroacetylacetonate. Tetrakis(pyridine)bis(trifluoroacetato)copper(II)," *Inorganic Chemistry* 18, no. 12 (1979): 3519–3522.

Prigogine, Ilya. *The End of Certainty.* New York: Free Press, 1997.

———. *Is Future Given?* River Edge, NJ: World Scientific Publishing, 2003.

———. *Las leyes del caos.* Editorial Crítica, 1999.

———. *¿Tan sólo una ilusión? Una exploración del caos al orden.* Tusquets editores, 1997.

Revel, J. F. *La gran mascarada, ensayo sobre la supervivencia de la utopía socialista.* Editorial Taurus, 2000.

———. *La tentación totalitaria.* Plaza & Janés Ed., 1976.

Revel, J. F., and M. Ricard. *El monje y el filósofo.* Edc. Urano, 1998.

Ricard, M., and T. X. Thuan. *El infinito en la palma de la mano.* Edc. Urano, 2001.

Rindler, Wolfgang. *Special Relativity.* London: Oliver and Boyd, 1960.

Russell, Bertrand. *The ABC of Relativity.* Great Britain: George Allen & Unwin, 1985.

———. *The Art of Philosophizing & Other Essays.* Totowa, NJ: Littlefield Adams, 1968.

———. *A Critical Exposition of the Philosophy of Leibnitz.* Cambridge: Cambridge University Press, 1900.

———. *A History of Western Philosophy.* New York: Simon and Schuster, 1945.

———. *Human Knowledge: Its Scope and Limits.* Great Britain: George Allen & Unwin, 1948.

———. *In Praise of Idleness.* Great Britain: George Allen & Unwin, 1960.

———. *Living Philosophies.* Creative Education, 1985.

———. *My Philosophical Development.* London: Routledge, 1995.

———. *An Outline of Philosophy.* London: Routledge, 1995.

———. *The Philosophy of Logical Atomism.* Chicago: Open Court Publishing, 1996.

———. *The Problems of Philosophy.* London: Williams and Norgate, 1912.

Schacter, Daniel. Searching for Memory, the Brain, the Mind, and the Past. New York: Basic Books, 1996.

Schlesinger, G. N. *Metaphysics.* New York: Barnes & Noble Books, 1983.

Schrödinger, Erwin. *What Is Life?* Cambridge: Cambridge University Press, 1988.

Searle, J. R. *The Rediscovery of the Mind.* Cambridge: MIT Press, 1992.

Serres, M., ed. *Historia de las ciencias.* Ediciones Cátedra, 1998.

Smith, John M., and Eörs Szathmáry. *The Origins of Life: From the Birth of Life to the Origin of Language.* New York: Oxford University Press, 1999.

Smith, Peter. *Explaining Chaos.* Cambridge: Cambridge University Press, 1999.

Strogatz, Steven H. *Nonlinear Dynamics and Chaos: With Applications to Physics, Biology, Chemistry, and Engineering.* New York: Perseus, 1994.

Szekely, E. B. *La enseñanza de los Esenios desde Enoch hasta los rollos del Mar Muerto.* Editorial Sirio, 1981.

Varela, F. *El fenómeno de la vida.* Dolmen Ediciones, 2002.

Watts, Alan. *The Philosophies of Asia.* North Clarendon, VT: Tuttle Publishing, 1995.

———. *Taoism: Way Beyond Seeking.* North Clarendon, VT: Tuttle Publishing, 1998.

———. *The Tao of Philosophy.* North Clarendon, VT: Tuttle Publishing, 1995.

Weaver, Jefferson H. *The World of Physics.* New York: Simon & Schuster, 1987.

Wilber, Ken. *The Atman Project: A Transpersonal View of Human Development.* Theosophical Publishing House, 1980.

Wittgenstein, Ludwig. *The Blue and Brown Books.* New York: Harper & Row, 1958.

———. *Philosophical Investigations.* Blackwell Publishing, 2001.

———. *Tractatus Logico Philosophicus.* Routledge Classics, 2001.

Zeilinger, Anton. "A Foundational Principle for Quantum Mechanics." *Foundations of Physics* 29, no. 4 (1999).

Zohar, Danah, and Dr. Ian Marshall. *SQ: Spiritual Intelligence.* New York: Bloomsbury Publishing, 2000.

ABOUT THE AUTHOR

Jaime Pradilla-Sorzano is a theoretical and experimental chemist and physicist who has dedicated his career to the advancement of science and applied science. He holds a Ph.D. in chemistry and a master's degree from Case Western Reserve University in Cleveland, Ohio, as well as BSc degrees in both chemistry and chemical engineering from the I.Q.S. in Barcelona.

Jaime has published many scientific and technical papers and articles in international and Colombian publications, most notably his dissertation on the Jahn-Teller effect in electron paramagnetic resonance of mono-crystals of copper compounds, published in cooperation with Professor and Distinguished Professor Emeritus John P. Fackler, Jr., at Texas A&M University. The Jahn-Teller effect is one of the most fascinating phenomena in modern physics and chemistry and is of importance for current scientific studies and development of high-temperature superconductors, one of the most relevant and promising fields of modern-day scientific discovery.

The chemical industry has also benefited from Jaime's contributions. His technical papers on industrial lubricants and fatty acids, PVC extrusion chemistry, and agricultural products have been used to advance industry and its methods. In addition, his spectroscopic studies of solid samples such as phosphoric rock, asphalts, clays, zeolitic catalysts, and starches have been used to improve many industrial processes. He has served on many boards of important industrial companies and as technical consultant to a myriad of chemical industry leaders.

Jaime is a highly decorated scholar who has held many important positions in academia, including dean of science at the U.I.S. in Colombia S.A., among others. In addition, he holds a diverse number of honors, including a lifetime achievement award for the advancement of science in Colombia from the ACEACE and professor emeritus from the U.I.S.

He is a current member of leading scientific associations such as the Sigma Chi Society, the Colombian Society of Crystallography, and the Sarria Institute's Chemical Society in Spain.

ABOUT THE TRANSLATOR

Ricardo Pradilla is a professional engineer registered in New Jersey who has dedicated his career to applied science and construction. He has held many important positions with large construction companies in New Jersey, including serving as vice president and on the board of directors for a leading construction company, as well as president of a division dedicated to preconstruction services. Ricardo has worked on projects ranging up to $500 million and holds a BSc in civil engineering from the U.I.S. in Colombia S.A. and an engineering master's degree from Case Western Reserve University, specializing in construction engineering and management. He now owns and operates a successful construction company in New Jersey.

www.ingramcontent.com/pod-product-compliance
Lightning Source LLC
Chambersburg PA
CBHW051903170526
45168CB00001B/219